T0304680

PUMPING

HOW TO ORDER THIS BOOK

BY PHONE: 800-233-9936 or 717-291-5609, 8AM–5PM Eastern Time

BY FAX: 717-295-4538

BY MAIL: Order Department
Technomic Publishing Company, Inc.
851 New Holland Avenue, Box 3535
Lancaster, PA 17604, U.S.A.

BY CREDIT CARD: American Express, VISA, MasterCard

BY WWW SITE: http://www.techpub.com

FUNDAMENTALS *for the* WATER & WASTEWATER MAINTENANCE OPERATOR SERIES

PUMPING

FRANK R. SPELLMAN, Ph.D.
JOANNE DRINAN

CRC Press is an imprint of the
Taylor & Francis Group, an **informa** business

Pumping
a**TECHNOMIC**® publication

Technomic Publishing Company, Inc.
851 New Holland Avenue, Box 3535
Lancaster, Pennsylvania 17604 U.S.A.

Main entry under title:
Fundamentals for the Water and Wastewater Maintenance Operator Series: Pumping

A Technomic Publishing Company book
Bibliography: p.
Includes index p. 213

Library of Congress Catalog Card No. 00-111215
ISBN No. 1-58716-014-5

Contents

11 POSITIVE-DISPLACEMENT PUMPS

12 FINAL REVIEW EXAMINATION

APPENDIX A

APPENDIX B

Series Preface

Currently several books address broad areas of wastewater and waterworks operation. Persons seeking information for professional development in water and wastewater can locate study guides and also find materials on technical processes such as activated sludge, screening and coagulation. What have not been available until now are accessible treatments of each of the numerous specialty areas that operators must master to perform plant maintenance activities and at the same time to upgrade their knowledge and skills for higher levels of certification.

The *Fundamentals for the Water and Wastewater Maintenance Operator Series* is designed to meet the needs of operators who require essential background knowledge of subjects often overlooked or covered superficially in other sources. Written specifically for maintenance operators, the series comprises focused books designed to enhance knowledge and understanding.

Fundamentals for the Water and Wastewater Maintenance Operator Series covers over a dozen subjects in volumes that form stand-alone information guides or elements of a library of key topics. Areas to be presented in series volumes include: electricity, electronics, water hydraulics, water pumps, handtools, blueprint reading, piping systems, lubrication, and data collection.

Each volume in the series is written in a straightforward style without jargon or complex calculations. All are heavily illustrated and include extensive, clearly outlined sample problems. Self-check tests are found within every chapter, and a comprehensive examination concludes each book.

The series provides operators with the information required for improved job performance. Equally important, using key points, worked problems, and sample test questions, the series is designed to help operators answer questions and solve problems on certification and licensure examinations.

Preface

It has often been said that designing pumps is a science; maintaining them is an art. Our experience has shown that this statement is true. This book's purpose is not to make pump designers out of anyone. Instead, in *Pumping*, our intent is to provide the essential information necessary to assist the water/wastewater operator towards an understanding of pump operation fundamentals and applications, which is helpful in preparing for licensure examinations. The book discusses maintenance procedures, including preventative maintenance and troubleshooting.

Anyone currently employed in the water or wastewater treatment industry can tell you flat out that times are changing in the industry. Some of the changes are dramatic—privatization, for example. Others are subtle and affect plant maintenance activities in less profound, but important, ways. Consider just one small example of recent change. In the past, plant electric motor controllers contained many electro-mechanical control devices. Today, these antiquated devices have been replaced by more compact and efficient micro-electronic controllers, reducing electrical maintenance requirements. Other changes are more apparent. For instance, many treatment plants have replaced routine operator sampling functions with automatic samplers.

However, in water/wastewater work, we see one fundamental element that we predict will remain largely immune to change—the pump, a simple machine that creates and maintains the pulse of both water and wastewater operations.

Because the hydraulic pump is so perfectly suited to the tasks it performs, and because the principles that make the pump work are physically fundamental, the idea that any new device would ever replace the pump is difficult to imagine. The hydraulic pump is the workhorse of water/wastewater operations and is unlikely to be superceded any time soon.

However, one observation we make in facility after facility is puzzling. If the pump, the heart of water/wastewater treatment, is so basic, so widely accepted and needed throughout the industry, why is this critical machine so frequently neglected or abused until it breaks down? When the heart in the human body malfunctions, the entire body is affected. This is the same with pumps in critical applications; when they fail, it affects the entire water/wastewater process. Many treatment facilities are unable to meet their water treatment or distribution requirements and wastewater effluent limits for one of three reasons:

1. Untrained operation and maintenance staff
2. Poor plant maintenance
3. Improper plant design

This text, addresses the first two of these three problem causes. The book provides plant or maintenance operators with a basic knowledge of the principles of operation for each type of centrifugal and positive-displacement pump commonly associated with water/wastewater treatment, including the maintenance requirement of each, common operational problems, appropriate corrective actions, and specific maintenance procedures, such as packing and changing a mechanical seal.

This text is not meant to replace the hands-on training an inexperienced person would need to acquire before performing all the maintenance on centrifugal and positive-displacement pumps. However, the information contained in this text, when combined with in-plant experience in basic pump maintenance procedures, should achieve the desired result: better performance through preservation of plant flexibility and reduction in unscheduled shutdowns of critical pump facilities.

Pumping focuses on the very few important principles operators need for working with or around pumping systems and concentrates on the practical side of the subject, stressing its importance in treatment, collections, and distribution operations.

To assure correlation to modern practices and design, we present illustrative problems in commonly used pumping hydraulic terms and parameters (head, capacity, work power, etc.), and cover typical pumping types and ancillary equipment found in today's water/wastewater systems.

Each chapter ends with a Self-Test to help you evaluate your mastery of the concepts presented. Before going on to the next chapter, take the Self-Test, compare your answers to the key, and review the pertinent information for any problems you missed. If you miss many items, review the whole chapter. A comprehensive final exam can be found at the end of the text.

***Note:** This symbol (to the left) displayed in various locations throughout this manual indicates a point is especially important and should be studied carefully.*

This text is accessible to those who have no experience with pumping systems; however, an understanding of basic water hydraulic principles is a plus (see *Hydraulics* in this series). If you work through the text systematically, you will be surprised at how easily you acquire an understanding of the basics of pumping systems.

Acknowledgements

To water and wastewater operators everywhere.

Introduction

Pumping facilities are required wherever gravity can't be used to supply water to the distribution system under sufficient pressure to meet all service demands.

Regarding wastewater, pumps are used to lift or elevate the liquid from a lower elevation to an adequate height at which it can flow by gravity or overcome hydrostatic head. There are many pumping applications at a wastewater treatment facility. These applications include pumping of (1) raw or treated wastewater, (2) grit, (3) grease and floating solids, (4) dilute or well-thickened raw sludge, or digested sludge (sludge or supernatant return), and (5) dispensing of chemical solutions. Pumps and lift stations are used extensively in the collection system. Each of the various pumping applications is unique and requires specific design and pump selection considerations.

Where pumping is necessary, it accounts for most of the energy consumed in water supply and/or wastewater treatment operations.[1]

TOPICS

1.1 ARCHIMEDES' SCREW

At the top of the small list comprised of names associated with some of the greatest achievements in science and the arts are Aristotle, Michelangelo, Da Vinci, Newton, and Einstein. You may have noticed that one name has been left off this list—Archimedes. While Archimedes may be recognized as one of the greatest geniuses of all time, many are confused about what he actually did. As Stein points out, all we may well remember is, "Something about running naked out of his bath crying 'Eureka, Eureka.'"

We were just as uninformed until we began the research for this text. We were genuinely astonished at the magnitude and the sheer number of

[1]From *Water Transmission and Distribution*, 2nd ed., Denver: American Water Works Association, p. 357, 1996; Qasim, S. R., *Wastewater Treatment Plants: Planning, Design, and Operation.* Lancaster, PA: Technomic Publishing Co., Inc., p. 177, 1994.

Key Terms Used in This Chapter

NORIA — A water wheel with buckets attached to its rim used to raise water from a stream, especially for transferral to an irrigation trough

SCREW PUMP — Consists of a U-shaped channel, into which a rotating screw fits closely. The channel, angled at inclinations up to 45°, takes water from a lower level and literally "screws" the water from the lower to a higher level. The screw, of course, does not develop any pressure, as it is merely a conveyor.

Archimedes' scientific accomplishments and their profound impact on today's world.[2]

Contrary to appearances, the goal of this text is not to make Archimedes' most mathematically significant discoveries (of which there are so many) the main topic of our discussion. As the title, *Pumping,* conveys, this text covers those aspects concerned with the basic science of pumping and many of their artful applications in the world of water and wastewater treatment.

We begin with Archimedes because—for our purposes—Archimedes *is* the beginning. Moreover, Archimedes is included in our discussion of basic pumping both to enrich the user's experience in reading this text and to enlarge the reader's historical perspective.

Few engineered artifacts are as essential as pumps in the development of the culture that our western civilization enjoys. Such machines affect every facet of our daily lives. Even before the time of Archimedes (287 BC), ancient civilizations requiring irrigation and essential water supplies used crude forms of pumps that (with their design refinements) are still in use even today.

IMPORTANT

***Note:** Exactly how significant and useful pumps are to civilization can be appreciated when you consider that, of all the machines currently used, the pump is the second most frequently used device. Only use of the electric motor exceeds the use of the pump.*

[2]Stein, S., *Archimedes: What Did He Do Besides Cry Eureka?* Washington, DC: The Mathematical Association of America, p. ix, 1999.

Krutzsch[3] fittingly points out that "only the sail can contend with the pump for the title of the earliest invention for the conversion of natural energy to useful work, and it is doubtful that the sail takes precedence." In reality, because the sail is not a machine, we can state unequivocally that the pump stands "essentially unchallenged as the earliest form of machine which substituted natural energy for muscular effort in the fulfillment of man's needs."

As historical records differ among ancient civilizations (cultures), and as each culture commonly supplied solutions to individual problems, several names and forms of the earliest pumps are known. Some cultures described the earliest pumps as water wheels, Persian wheels, or norias (i.e., water wheels of various design; a noria is a water wheel with buckets attached to its rim that are used to raise water from a stream, especially for transferal to an irrigation trough). Even today, water wheels of similar design have continued in use in parts of the Orient.

Where does Archimedes come in? The Archimedean screw is probably the best known of the early pumps. In fact, the principle of the Archimedean screw is still being used today. Figure 1.1 shows a system application of Archimedes' screw lift pumps as applied in wastewater treatment.

Let's take an even closer look at Archimedes' invention (a modern view that includes modern applications).

As previously stated and as shown in Figure 1.1, Archimedean screw pumps are occasionally used for raw wastewater pumping applications. According to Benjes and Foster, these units are "advantageous in that they do not require a conventional wet well and they are self-compensating in that they automatically pump the liquid received regardless of quantity as long as it does not exceed the design capacity of the pump." In addition, no special drive equipment is required. Moreover, the total operating head of a screw pump installation is less than pumps that require conventional suction and discharge piping (see Figure 1.2). Screw pumps, however, are limited by pumping head and are not used for lifts more than 25 ft.[4]

[3]Krutzsch, W. C., Introduction and Classification of Pumps. In *Pump Handbook* by Karassik, I. J., et al. (eds.). New York: McGraw-Hill Book Company, p. 1, 1976.

[4]Benjes, H. H., Sr. and Foster, W. E. Sewage. In *Pump Handbook*, Karassik, I. J. et al. (eds.). New York: McGraw-Hill Book Company, pp. 10-29, 1976.

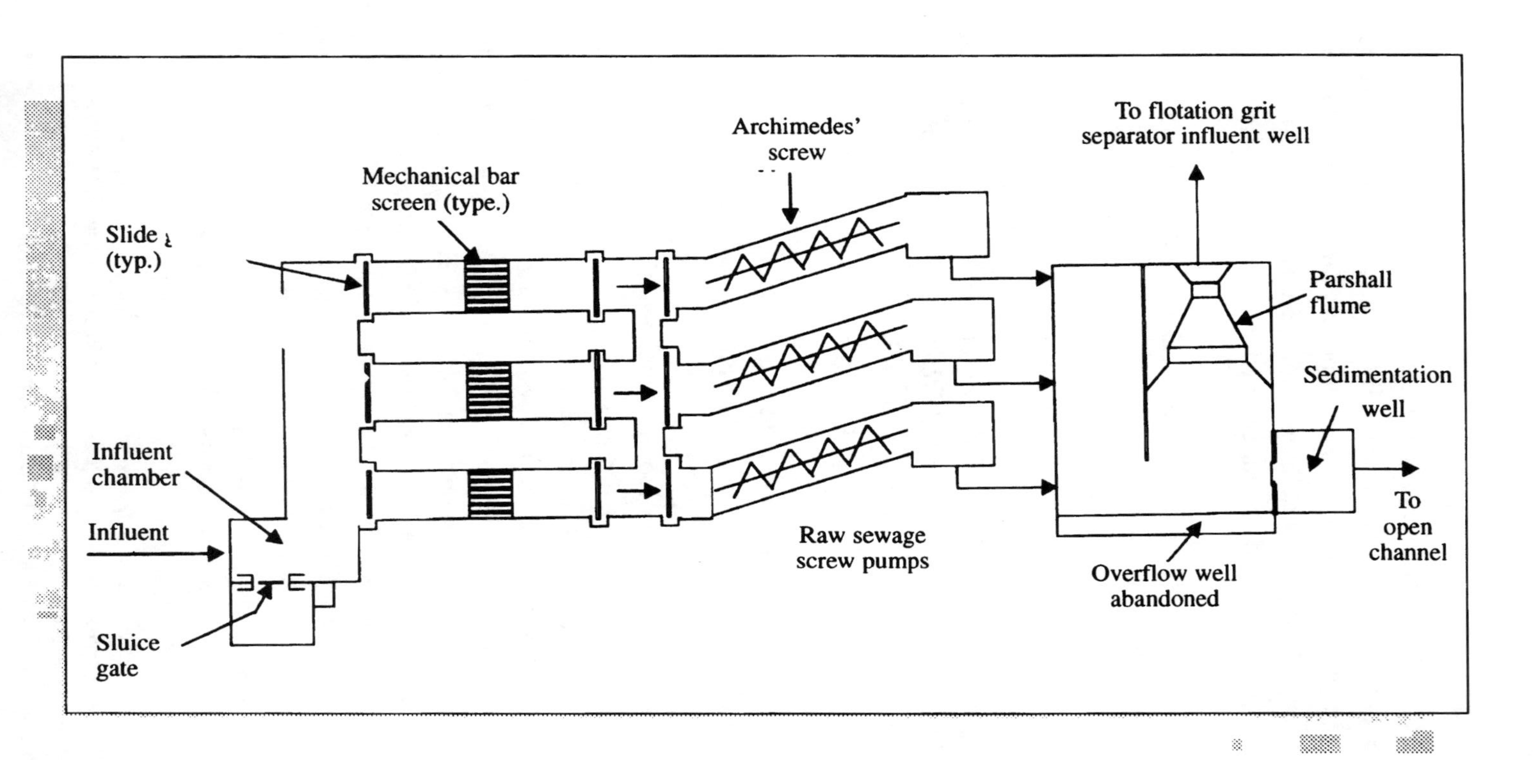

Figure 1.1
Archimedes' screw lift pumps as applied in wastewater treatment.

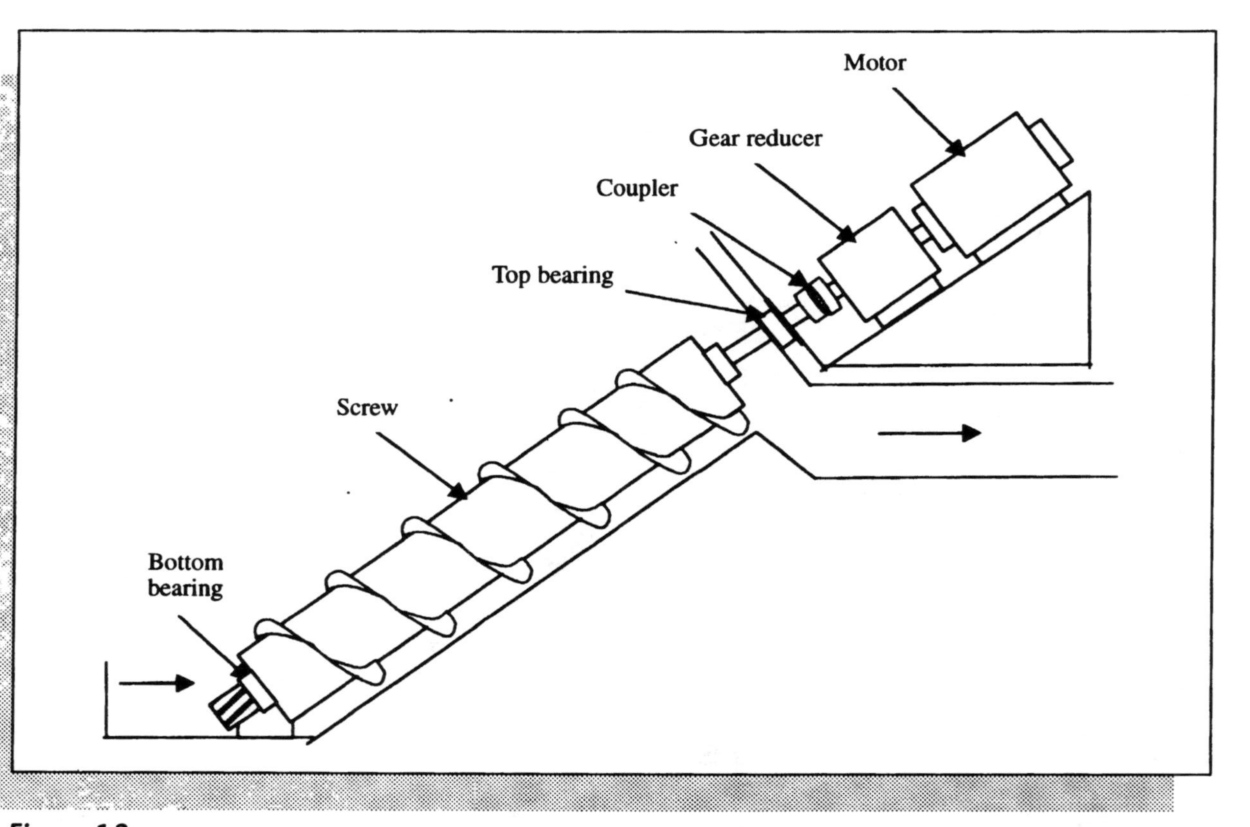

Figure 1.2
Archimedean screw pump.

IMPORTANT **Note:** *For more information about Archimedes and his remarkable discoveries and inventions, we highly recommend Sherman Stein's* Archimedes: What Did He Do Besides Cry Eureka? *available from the Mathematical Association of America, Washington, D.C., 1-800-331-1622.*

REFERENCES

Benjes, H. H., Sr. and Foster, W. E., Sewage. In *Pump Handbook.* Karassik, I. J. et al. (eds.). New York: McGraw-Hill Book Company, pp. 10–29, 1976.

Krutzsch, W. C., Introduction and Classification of Pumps. In *Pump Handbook.* Karassik, I. J. et al. (eds.). New York: McGraw-Hill Book Company, p. 1, 1976.

Qasim, S. R., *Wastewater Treatment Plants: Planning, Design, and Operation.* Lancaster, PA: Technomic Publishing Co., Inc., p. 177, 1994.

Stein, S., *Archimedes: What Did He Do Besides Cry Eureka?* Washington, D.C.: The Mathematical Association of America, p. ix, 1999.

Water Transmission and Distribution, 2nd ed., Denver: American Water Works Association, p. 357, 1996.

Self-Test

1.1 Why did Archimedes cry "Eureka, Eureka"? (Note: If you do not know the answer to this question, use an encyclopedia to find out.)

Basic Water Hydraulics

A water pumping system can be equated with the human circulatory system.

In human beings, the blood, kept in motion by the pumping of the heart, circulates through a series of vessels. The heart is actually a double pump: the right side pumps blood to the lungs and the left side pumps blood to the rest of the body.

In a pumping system, the fluid, kept in motion by the pumping of the pump (which is driven by a prime mover—usually an electric motor), circulates through a series of pipes, valves, and various other appurtenances.

Both of these hydraulic "machines," the heart and the pump, perform vital functions.

TOPICS

Definitions
Basic Principles of Water Hydraulics
Basic Pumping Calculations
Pump Characteristic Curves
Pumps in Series and Parallel
Considerations for Pumping Wastewater
Types of Pumps Used in Water/Wastewater Treatment

2.1 INTRODUCTION

Hydraulics is the study of fluids at rest and in motion. It is essential for an understanding of how water/wastewater systems work, especially water distribution and wastewater collection systems, both of which may rely heavily upon the motive force provided by pumps. Pumps, in turn, use energy to keep water and wastewater moving. To operate a pump efficiently, the operator and/or maintenance operator must be familiar with several basic principles of hydraulics.

Keep in mind that this chapter is not intended to provide a detailed discussion of the principles of hydraulics; it is rather a very brief discussion of the essential principles relating to the operation of pumps. Readers wishing to obtain additional information on the basic science of water hydraulics should consult the *Water Hydraulics* volume of this series.

2.2 DEFINITIONS AND KEY TERMS[5]

There are several basic terms and symbols used in discussing pumping hydraulics that should be known and understood by those who must operate and maintain plant pumping facilities. The most important terms are included in this section.

Absolute Pressure—This is the pressure of the atmosphere on a surface. At sea level, a pressure gauge with no external pressure added will read 0 psig. The atmospheric pressure is 14.7 psia. If the gauge reads 15 psig, the absolute pressure will be 15 + 14.7, or 29.7 psia.

Acceleration Due to Gravity (*g*)—The rate at which a falling body gains speed. The acceleration due to gravity is 32 feet/second/second. This simply means that a falling body or fluid will increase the speed at which it is falling by 32 feet/second every second that it continues to fall.

Affinity Laws—Any machine that imparts velocity and converts a velocity to pressure can be categorized by a set of relationships that apply to any dynamic conditions. These relationships are referred to as the "Affinity Laws." They can be described as similarity processes, that follow the following rules:

1. Capacity varies as the rotating speed—that is, the peripheral velocity of the impeller
2. Head varies as the square of the rotating speed
3. Bhp varies as the cube of the rotating speed

Atmospheric Pressure—The pressure exerted on a surface area by the weight of the atmosphere is atmospheric pressure, which at sea level is 14.7 psi, or one atmosphere. At higher altitudes, the atmospheric pres-

[5]From Lindeburg, M. R., *Civil Engineering Reference Guide,* 4th ed. San Carlos, CA: Professional Publications, pp. 4.2–4.25, 1986; Wahren, U., *Practical Introduction to Pumping Technology.* Houston: Gulf Publishing Company, pp. 3–8, 1997; Garay, P. N., *Pump Application Book.* Lilburn, GA: The Fairmont Press, pp. 23, 143, 162, 1990.

sure decreases. At locations below sea level, the atmospheric pressure rises (see Table 2.1).

Cavitation—An implosion of vapor bubbles in a liquid inside a pump caused by a rapid local pressure decrease occurring mostly close to or touching the pump casing or impeller. As the pressure reduction continues, these bubbles collapse or implode. Cavitation may produce noises that sound like pebbles rattling inside the pump casing and may also cause the pump to vibrate and to lose hydrodynamic efficiency. This effect contrasts boiling, which happens when heat builds up inside the pump.

Continued serious cavitation may destroy even the hardest surfaces. Avoiding cavitation is one of the most important pump design criteria. Cavitation limits the upper and lower pump sizes, as well as the pump's peripheral impeller speed.

Cavitation may be caused by any of the following conditions:

- Discharge heads far below the pump's calibrated head at peak efficiency
- Suction lift higher or suction head lower than the manufacturer's recommendation
- Speeds higher than the manufacturer's recommendation
- Liquid temperatures (thus, vapor pressure) higher than that for which the system was designed

Critical Speed—At this speed, a pump may vibrate enough to cause damage. Pump manufacturers try to design pumps with the first critical

TABLE 2.1. Atmospheric Pressure at Various Altitudes.

Altitude	Barometric Pressure	Equivalent Head
−1000 ft	15.2 psi	35.2 ft
Sea Level	14.7psi	34.0 ft
1500 ft	13.9 psi	32.2 ft
3000 ft	13.2 psi	30.5 ft
5000 ft	12.2 psi	28.3 ft
7000 ft	11.3 psi	26.2 ft
8000 ft	10.9 psi	25.2 ft

speed at least 20% higher or lower than rated speed. Second and third critical speeds usually don't apply in pump usage.

Cross-sectional Area (*A*)—The area perpendicular to the flow that the load in a channel or pipe occupies (see Figure 2.1).

Density—The mass per unit volume measured in pounds per cubic foot at 68°F or in grams per milliliter at 4°C.

Discharge Pressure—The pressure measured at the pump's discharge nozzle. Measurements may be stated in psig, bars, kg/cm², or kilopascals.

Displacement—The capacity, or flow, of a pump. This measurement, primarily used in connection with positive displacement pumps, is measured in units such as gallons, cubic inches, and liters.

Energy—The ability to do work.

- *Potential Energy*—energy due to the liquid's location or condition
- *Kinetic Energy*—energy of motion

Flow—The volume or amount of a liquid moving through a channel or pipe. It is measured in million gallons/day, gallons/day, and cubic feet/second. In most hydraulic calculations, the flow is expressed in cubic feet/second, cfs. To obtain cubic feet per second when flow is given in million gallons per day, multiply by 1.55 cfs/MGD.

$$Q\text{, cfs} = \text{MGD} \times 1.55 \text{ cfs/MGD} \quad (2.1)$$

Head—The energy a liquid possesses at a given point or a pump must supply to move a liquid to a given location. Head is expressed in feet. Of

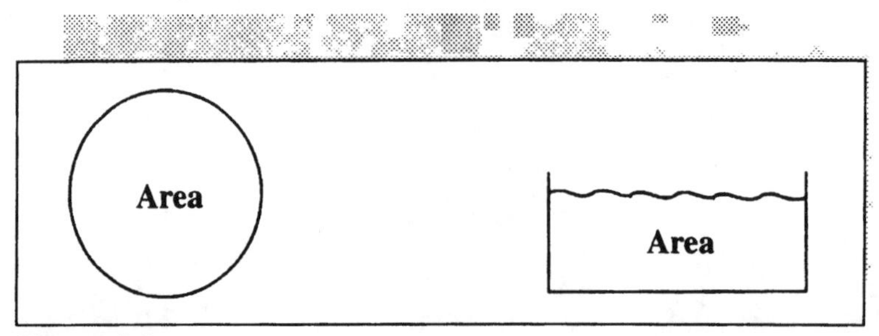

Figure 2.1
Cross-sectional area.

course, any head term can be converted to pressure by using Equation (2.2).

$$p = \rho \times h \tag{2.2}$$

where

p = pressure, psf
r = density, lb/ft3
h = head, ft

- *Cut Off Head*—The head at which the energy supplied by a pump and the energy required to move the liquid to a specified point are equal and no discharge at the desired point will occur.
- *Friction Head*—The amount of energy in feet that is necessary to overcome the resistance of flow that occurs in the pipes and fixtures (i.e., fittings, valves, entrances, and exits) through which the liquid is flowing.
- *Pressure Head*—The vertical distance a liquid can be raised by a given pressure. For example, if a liquid has a pressure of 1 pound per square inch (psi), the liquid will rise to a height of 2.31 feet.
- *Pump Head*—The energy in feet that a pump supplies to the fluid.
- *Static Head*—The energy in feet required to move a fluid from the supply tank to the discharge point (see Figure 2.2).
- *Total Head*—The total energy in feet required to move a liquid from the supply tank to the discharge point, taking into account the velocity head and the friction head (see Figures 2.3 and 2.4).
- *Velocity Head*—The energy in feet required to maintain a given speed in the liquid being moved. If the pump inlet nozzle and discharge nozzle are of equal size, then this term is normally zero.

$$\textbf{Velocity Head } (h_v) = V^2/2g \tag{2.3}$$

where

V = liquid velocity in a pipe

g = gravity acceleration, influenced by both altitude and latitude. At sea level and 45° latitude, it is 32.17 ft/sec/sec.

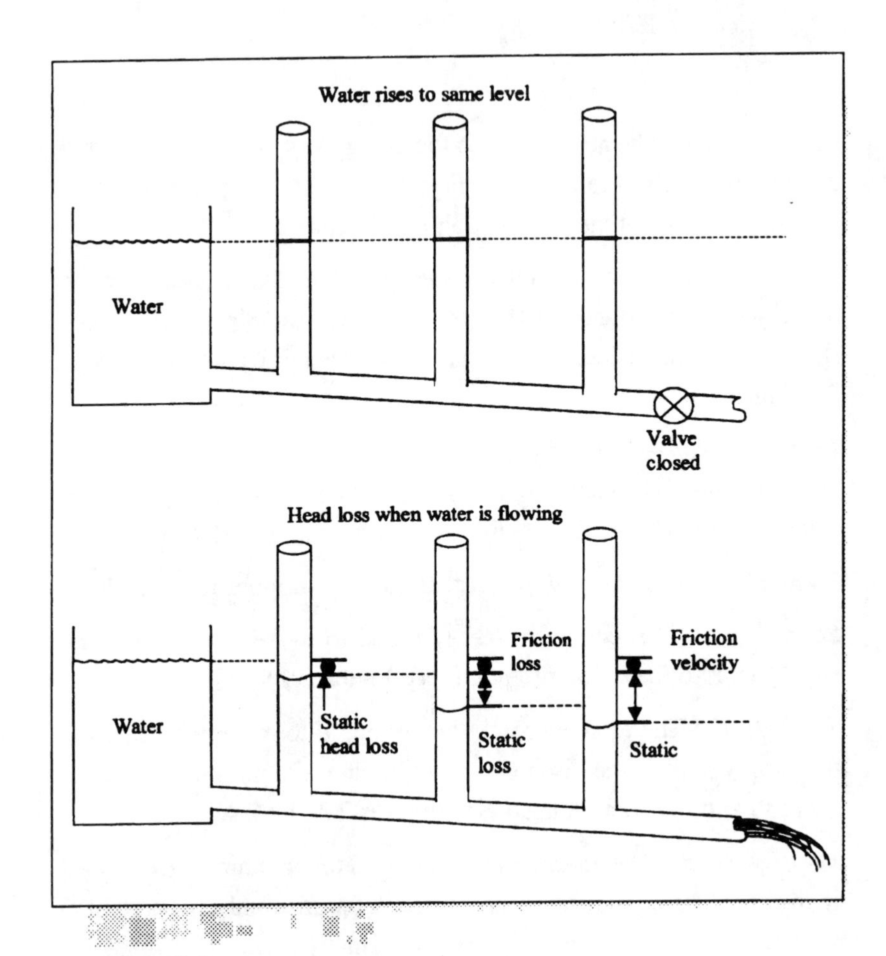

Figure 2.2
Head loss in non-pumping system.

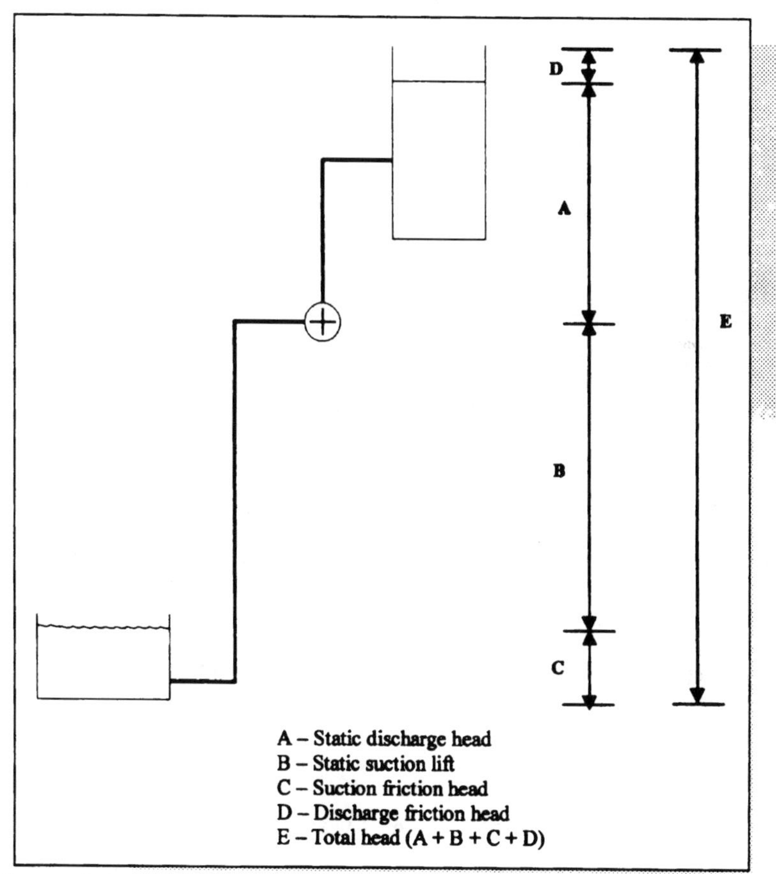

Figure 2.3
Head components for suction lift system.

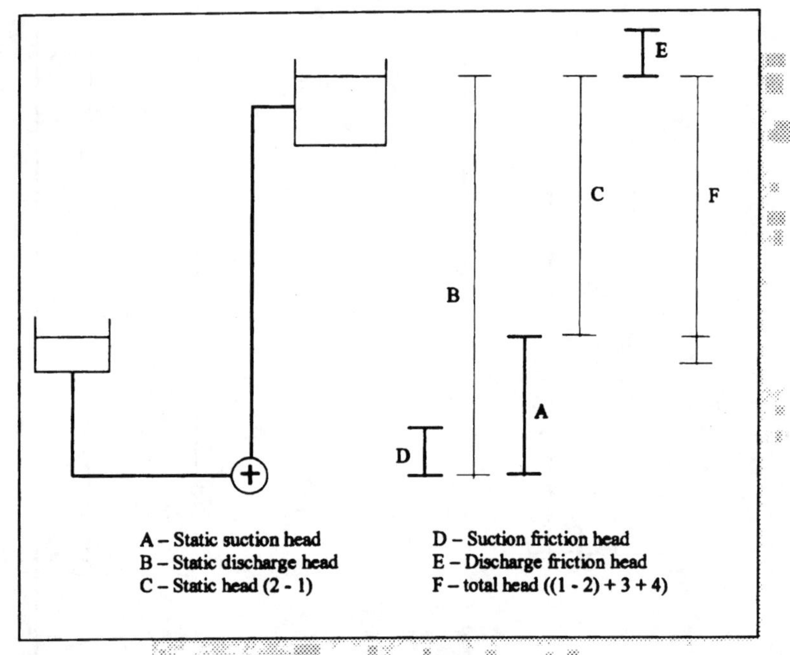

Figure 2.4
Head components for suction head type system.

- *Suction Head*—The total head in feet on the suction or supply side of the ɪ ump when the supply is located above the center of the pump.
- *Discharge Head*—The total head in feet on the discharge side of the pump.
- *Suction Lift*—The total head in feet on the suction or supply side of the pump, when the supply is located below the center of the pump.
- *Total Differential Head*—The difference between the discharge head and the suction head, expressed in feet or meters.

Horsepower (hp)—Work a pump performs while moving a determined amount of liquid at a given pressure.

- *Hydraulic (Water) Horsepower (whp)*—Pump output is measured in whp.
- *Brake Horsepower*—The input horsepower delivered to the pump shaft.

Minimum Flow—the lowest continuous flow at which a manufacturer will guarantee a pump's performance.

Minimum Flow Bypass—A pipe leading from the pump discharge piping back into the pump suction system. A pressure control or flow control valve opens this line when the pump discharge flow approaches the pump's minimum flow value. The purpose is to protect the pump from damage.

Net Positive Suction Head (NPSH)—The net positive suction head (NPSH) available is the NPSH in feet available at the centerline of the pump inlet flange. The NPSH required (NPSHR) refers to the NPSH specified by a pump manufacturer for proper pump operation.

Power—Use of energy to perform a given amount of work in a specified length of time. In most cases, this is expressed in terms of horsepower.

Pressure—A force applied to a surface. The measurements for pressure can be expressed as various functions of psi, or pounds per square inch, such as

- atmospheric pressure (psi) = 14.7 psi
- metric atmosphere = psi × 0.07
- kilograms per square centimeter (kg/cm^2) = psi × 0.07
- kilopascals = psi × 6.89
- bars = psi × 14.50

Pump Performance Curves—Performance curves for centrifugal pumps are different in kind from curves drawn for positive displacement pumps. This is because the centrifugal pump is a dynamic device, in that the performance of the pump responds to forces of acceleration and velocity. Note that every specific performance curve is based on a particular speed and a specific impeller diameter, impeller width, and fluid viscosity (usually taken as the viscosity of water).

Specific Gravity—The result of dividing the weight of an equal volume of water at 68°F. If the data are in grams per milliliter, the specific gravity of a body of water is the same as its density at 4°C.

Specific Speed—In the case of centrifugal pumps, a correlation of pump capacity, head, and speed at optimum efficiency is used to classify the pump impellers with respect to their specific geometry. This correlation is called specific speed, and is an important parameter for analyzing pump performance.

Suction Pressure—The pressure, in psig, at the suction nozzle's centerline.

Vacuum—Any pressure below atmospheric pressure is a partial vacuum. The expression for vacuum is in inches or millimeters of mercury (Hg). Full vacuum is at 30 in. Hg. To convert inches to millimeters, multiply inches by 25.4.

Vapor Pressure (vp)—At a specific temperature and pressure, a liquid will boil. The point at which the liquid begins to boil is the liquid's vapor pressure. The vapor pressure will vary with changes in either temperature or pressure, or both.

Velocity (*V*)—The speed of the fluid moving through a pipe or channel. It is normally expressed in feet per second (fps).

Volumetric Efficiency—Found by dividing a pump's actual capacity by the calculated displacement. The expression is primarily used in connection with positive displacement pumps.

Work—Using energy to move an object a distance. It is usually expressed in foot-pounds.

2.3 BASIC PRINCIPLES OF WATER HYDRAULICS[6]

Recall that *hydraulics* is defined as the study of fluids at rest and in motion. While basic principles apply to all fluids, for our purposes, we

[6]From notes taken at *The Waterworks Operators Short Course*. Blacksburg, VA: Virginia Tech, August 1999.

consider only those principles that apply to water/wastewater. [Note: Although much of the basic information that follows is concerned with the hydraulics of distribution systems (i.e., piping), it is important for the operator to understand these basics in order to more fully appreciate the function of pumps.]

2.3.1 WEIGHT OF AIR

Our study of water hydraulics begins with air! A blanket of air, many miles thick, surrounds the earth. The weight of this blanket on a given square inch of the earth's surface will vary according to the thickness of the atmospheric blanket above that point. At sea level, the pressure exerted is 14.7 pounds per square inch (psi). On a mountaintop, air pressure decreases because the blanket is not as thick.

2.3.2 WEIGHT OF WATER

Because water must be stored and moving in water supplies and wastewater must be collected, processed in unit processes, and outfalled to its receiving body, we must consider some basic relationships in the weight of water. One cubic foot of water weighs 62.4 pounds and contains 7.48 gallons. One cubic inch of water weighs 0.0362 pounds. Water one foot deep will exert a pressure of 0.43 pounds per square inch on the bottom area (12 in x 0.062 lb/in.^{3}). A column of water two feet high exerts 0.86 psi, one 10 feet high exerts 4.3 psi, and one 52 feet high exerts

$$52 \text{ ft} \times 0.43 \text{ psi/ft} = 22.36 \text{ psi}$$

A column of water 2.31 feet high will exert 1.0 psi. To produce a pressure of 40 psi requires a water column

$$40 \text{ psi} \times 2.31 \text{ ft/psi} = 92.4$$

The term *head* is used to designate water pressure in terms of the height of a column of water in feet. For example, a 10-foot column of water exerts 4.3 psi. This can be called 4.3-psi pressure or 10 feet of head.

Another example: If the static pressure in a pipe leading from an elevated water storage tank is 37 pounds per square inch (psi), what is the elevation of the water above the pressure gauge?

Remembering that 1 psi = 2.31 and that the pressure at the gauge is 37 psi,

$$37 \text{ psi} \times 2.31 \text{ ft/psi} = 85.5 \text{ ft} \quad \text{(rounded)}$$

2.3.3 WEIGHT OF WATER RELATED TO THE WEIGHT OF AIR

The theoretical atmospheric pressure at sea level (14.7 psi) will support a column of water 34 feet high:

$$14.7 \text{ psi} \times 2.31 \text{ ft/psi} = 33.957 \text{ ft or } 34 \text{ ft}$$

At an elevation of one mile above sea level, where the atmospheric pressure is 12 psi, the column of water would be only 28 feet high (psi 12 × 2.31 ft/psi = 27.72 ft or 28 ft).

If a tube is placed in a body of water at sea level (a glass, a bucket, a water storage reservoir, or a lake, pool, etc.), water will rise in the tube to the same height as the water outside the tube. The atmospheric pressure of 14.7 psi will push down equally on the water surface inside and outside the tube.

However, if the top of the tube is tightly capped and all of the air is removed from the sealed tube above the water surface, forming a *perfect vacuum*, the pressure on the water surface inside the tube will be 0 psi. The atmospheric pressure of 14.7 psi on the outside of the tube will push the water up into the tube until the weight of the water exerts the same 14.7 psi pressure at a point in the tube even with the water surface outside the tube. The water will rise 14.7 psi × 2.31 ft/psi = 34 feet.

In practice, it is impossible to create a perfect vacuum, so the water will rise somewhat less than 34 feet; the distance it rises depends on the amount of vacuum created.

EXAMPLE 2.1

Problem: If enough air was removed from the tube to produce an air pressure of 9.7 psi above the water in the tube, how far will the water rise in the tube?

Solution: To maintain the 14.7 psi at the outside water surface level, the water in the tube must produce a pressure of 14.7 psi – 9.7 = 5.0 psi. The height of the column of water that will produce 5.0 psi is

$$5.0 \text{ psi} \times 2.31 \text{ ft/psi} = 11.5 \text{ ft} \quad \text{(rounded)}$$

2.3.4 WATER AT REST

Stevin's Law states, "The pressure at any point in a fluid at rest depends on the distance measured vertically to the free surface and the density of the fluid." Stated as a formula, this becomes

$$p = w \times h \qquad (2.4)$$

where

p = pressure in pounds per square foot (psf)
w = density in pounds per cubic foot (lb/ft3)
h = vertical distance in feet

EXAMPLE 2.2

Problem: What is the pressure at a point 15 feet below the surface of a reservoir?

Solution: To calculate this, we must know that the density of water, w, is 62.4 pounds per cubic foot. Thus,

$$p = w \times h$$
$$= 62.4 \text{ lb/ft}^3 \times 15 \text{ ft}$$
$$= 936 \text{ lb/ft}^2 \text{ or } 936 \text{ psf}$$

Waterworks/wastewater operators generally measure pressure in pounds per square **inch** rather than pounds per square **foot**; to convert, divide by 144 $in.^2/ft^2$ (12 in. x 12 in. = 144 $in.^2$):

$$p = \frac{936 \text{ lb/ft}^2}{144 \text{ in.}^2/\text{ft}^2} = 6.5 \text{ lb/in.}^2 \text{ or psi}$$

2.3.5 GAUGE PRESSURE

We defined *head* as the height a column of water would rise due to the pressure at its base. We demonstrated that a perfect vacuum plus atmospheric pressure of 14.7 psi would lift the water 34 feet. If we now open the top of the sealed tube to the atmosphere and enclose the reservoir, then increase the pressure in the reservoir, the water will again rise in the tube. Because atmospheric pressure is essentially universal, we usually ignore the first 14.7 psi of actual pressure measurements and measure only the difference between the water pressure and the atmospheric pressure; we call this *gauge pressure.*

EXAMPLE 2.3

Problem: Water in an open reservoir is subjected to the 14.7 psi of atmospheric pressure, but subtracting this 14.7 psi leaves a gauge pressure of 0 psi. This shows that the water would rise 0 feet above the reservoir surface. If the gauge pressure in a water main is 100 psi, how far would the water rise in a tube connected to the main?

Solution:

$$100 \text{ psi} \times 2.31 \text{ ft/psi} = 231 \text{ ft}$$

2.3.6 WATER IN MOTION

The study of water flow is much more complicated than that of water at rest. It is important to have an understanding of these principles because the water/wastewater in a treatment plant and/or distribution/collection system is nearly always in motion (much of this motion is the result of pumping, of course).

2.3.6.1 DISCHARGE

Discharge is the quantity of water passing a given point in a pipe or channel during a given period of time. It can be calculated by the formula:

$$Q = V \times A \qquad (2.5)$$

where

Q = discharge in cubic feet per second (cfs)
V = water velocity in feet per second (fps or ft/sec)
A = cross-section area of the pipe or channel in square feet (ft^2)

The discharge can be converted from cfs to other units such as gallons per minute (gpm) or million gallons per day (MGD) by using appropriate conversion factors.

EXAMPLE 2.4

Problem: A pipe 12 inches in diameter has water flowing through it at 10 feet per second. What is the discharge in (a) cfs, (b) gpm, and (c) MGD?

Solution: Before we can use the basic formula, we must determine the area (A) of the pipe. The formula for the area is

$$A = \pi \times \frac{D^2}{4} = \pi \times r^2 \qquad (2.6)$$

(π is the constant value 3.14159)

where

D = diameter of the circle in feet
r = radius of the circle in feet

So, the area of the pipe is

$$A = \pi \times \frac{D^2}{4} = 3.14159 \times \frac{1^2}{4} = 0.785 \text{ ft}^2$$

Now, we can determine the discharge in cfs [part (a)]:

$$Q = V \times A = 10 \text{ ft/sec} \times 0.785 \text{ ft}^2$$
$$= \mathbf{7.85 \text{ ft}^3\text{/sec or cfs}}$$

For part (b), we need to know that 1 cubic foot per second is 449 gallons per minute, so 7.85 cfs × 449 gpm/cfs = **3520 gpm.**

Finally, for part (c), 1 million gallons per day is 1.55 cfs, so

$$\frac{7.85 \text{ cfs}}{1.55 \text{ cfs/MGD}} = \mathbf{5.06 \text{ MGD}}$$

2.3.6.2 THE LAW OF CONTINUITY

The Law of Continuity states that the discharge at each point in a pipe or channel is the same as the discharge at any other point (provided water does not leave or enter the pipe or channel). In equation form, this becomes

$$Q_1 = Q_2 \text{ or } A_1V_1 = A_2V_2 \qquad (2.7)$$

EXAMPLE 2.5

Problem: A pipe 12 inches in diameter is connected to a 6-inch diameter pipe. The velocity of the water in the 12-inch pipe is 3 fps. What is the velocity in the 6-in. pipe? Using the equation $A_1V_1 = A_2V_2$, determine the area of each pipe.

12-inch pipe: $A = \pi \times \frac{D^2}{4}$

$= 3.14159 \times \frac{(1 \text{ ft})^2}{4}$

$= 0.785 \text{ ft}^2$

6-inch pipe: $A = 3.14159 \times \frac{(0.5)^2}{4}$

$= 0.196 \text{ ft}^2$

The continuity equation now becomes

$$(0.785 \text{ ft}^2) \times (3 \text{ ft/sec}) = (0.196 \text{ ft}^2) \times V_2$$

Solving for V_2,

$$V_2 = \frac{(0.785 \text{ ft}^2) \times (3 \text{ ft/sec})}{(0.196 \text{ ft}^2)}$$

$$= 12 \text{ ft/sec or fps}$$

2.3.7 PIPE FRICTION

The flow of water in pipes is caused by the pressure applied behind it either by gravity or by hydraulic machines (pumps). The flow is retarded by the friction of the water against the inside of the pipe. The resistance of flow offered by this friction depends on the size (diameter) of the pipe, the roughness of the pipe wall, and the number and type of fittings (bends, valves, etc.) along the pipe. It also depends on the speed of the water through the pipe—the more water you try to pump through a pipe, the more pressure it will take to overcome the friction. The resistance can be expressed in terms of the additional pressure needed to push the water through the pipe, in either psi or feet of head. Because it is a reduction in pressure, it is often referred to as *friction loss* or *head loss.*

Friction loss increases as

- flow rate increases
- pipe diameter decreases
- pipe interior becomes rougher
- pipe length increases
- pipe is constricted
- bends, fittings, and valves are added

The actual calculation of friction loss is beyond the scope of this text. Many published tables give the friction loss in different types and diameters of pipe and standard fittings. What is more important here is recognition of the loss of pressure or head due to the friction of water flowing through a pipe.

One of the factors in friction loss is the roughness of the pipe wall. A number called the *C* factor indicates pipe wall roughness; the **higher** the *C* factor, the **smoother** the pipe.

Note: *C factor is derived from the letter C in the Hazen-Williams equation for calculating water flow through a pipe.*

Some of the roughness in the pipe will be due to the material; cast iron pipe will be rougher than plastic, for example. Additionally, the roughness will increase with corrosion of the pipe material and deposits of sediments in the pipe. New water pipes should have a *C* factor of 100 or more; older pipes can have *C* factors much lower than this.

In determining *C* factor, published tables are usually used. In addition, when the friction losses for fittings are factored in, other published tables are available to make the proper determinations. It is standard practice to calculate the head loss from fittings by substituting the *equivalent length of pipe,* which is also available from published tables.

2.4 BASIC PUMPING CALCULATIONS[7]

Certain computations used for determining various pumping parameters are important to the water/wastewater operator.

2.4.1 PUMPING RATES

Important Point: *The rate of flow produced by a pump is expressed as the volume of water pumped during a given period.*

The mathematical problems most often encountered by water/wastewater operators in regards to determining pumping rates are often deter-

[7]From Wahren, U., *Practical Introduction to Pumping Technology.* Houston: Gulf Publishing Company, pp. 10–17, 1997.

mined by using Equations (2.8) and/or (2.9).

Pumping Rate, (gpm) = gallons/minutes *(2.8)*

Pumping Rate, (gph) = gallons/hours *(2.9)*

EXAMPLE 2.6

Problem: The meter on the discharge side of the pump reads in hundreds of gallons. If the meter shows a reading of 110 at 2:00 p.m. and 320 at 2:30 p.m., what is the pumping rate expressed in gallons per minute?

Solution: The problem asks for pumping rate in **gallons per minute (gpm),** so we use Equation (2.8).

$$\text{Pumping rate, gpm} = \text{gallons/minutes}$$

Step 1: To solve this problem, first find the total gallons pumped (determined from the meter readings):

$$\begin{array}{r} 32{,}000 \text{ gallons} \\ -\ 11{,}000 \text{ gallons} \\ \hline 21{,}000 \text{ gallons} \end{array}$$

Step 2: The volume was pumped between 2:00 p.m. and 2:30 p.m., for a total of 30 minutes. From this information, calculate the gallons-per-minute pumping rate:

$$\text{Pumping rate, gpm} = \frac{21{,}000 \text{ gal}}{30 \text{ min}}$$

$$= 700 \text{ gpm pumping rate}$$

EXAMPLE 2.7

Problem: During a 15-minute pumping test, 16,400 gallons were pumped into an empty rectangular tank. What is the pumping rate in gallons per minute?

Solution: The problem asks for the pumping rate in gallons per minute, so again, we use Equation (2.8)

$$\text{Pumping rate, gpm} = \frac{\text{gallons}}{\text{minutes}}$$

$$= \frac{16{,}400 \text{ gallons}}{15 \text{ minutes}}$$

$$= 1093 \text{ gpm pumping rate (rounded)}$$

EXAMPLE 2.8

Problem: A tank 50 feet in diameter is filled with water to a depth of 4 feet. To conduct a pumping test, the outlet valve to the tank is closed, and the pump is allowed to discharge into the tank. After 80 minutes, the water level is 5.5 feet. What is the pumping rate in gallons per minute?

Solution:

Step 1: We must first determine the volume pumped in cubic feet:

$$\text{Volume pumped} = (\text{area of circle})(\text{depth})$$
$$= (0.785)(50\text{ ft})(50\text{ ft})(1.5\text{ ft})$$
$$= 2944\text{ ft}^3\text{ (rounded)}$$

Step 2: Convert the cubic-feet volume to gallons:

$$(2944\text{ ft}^3)(7.48\text{ gal/ft}^3)$$
$$= 22{,}021\text{ gallons (rounded)}$$

Step 3: The pumping test was conducted over a period of 80 minutes. Using Equation (2.8), calculate the pumping rate in gallons per minute:

$$\text{pumping rate} = \frac{\text{gallons}}{\text{minutes}}$$

$$= \frac{22{,}021\text{ gallons}}{80\text{ minutes}}$$

$$= 275.3\text{ gpm (rounded)}$$

2.4.2 CALCULATING HEAD LOSS

IMPORTANT

Important Note: *Pump head measurements are used to determine the amount of energy a pump can or must impart to the water; they are measured in feet.*

One of the principal calculations used in pumping problems is determining *head loss*. The following formula is used to calculate head loss:

$$H_f = K(V^2/2g)$$

where

H_f = friction head
K = friction coefficient
V = velocity in pipe
g = gravity (32.17 ft/sec/sec)

2.4.3 CALCULATING HEAD

For centrifugal pumps and positive displacement pumps, several other important formulae are used in determining *head.* In centrifugal pump calculations, the conversion of the discharge pressure to discharge head is the norm. Positive displacement pump calculations often leave given pressures in psi.

In the following formulae, *W* expresses the specific weight of liquid in pounds per cubic foot. For water at 68°F, *W* is 62.4 lb/ft^3. A water column 2.31 feet high exerts a pressure of 1 psi on 64°F water. Use the following formulae to convert discharge pressure in psig to head in feet:

- centrifugal pumps

$$H\text{, ft} = \frac{P\text{, psig} \times 2.31}{\text{specific gravity}} \tag{2.10}$$

- positive displacement pumps

$$H\text{, ft} = \frac{P\text{, psig} \times 144}{W} \tag{2.11}$$

To convert head into pressure:

- centrifugal pumps

$$P\text{, psi} = \frac{H\text{, ft} \times \text{specific gravity}}{2.31} \tag{2.12}$$

- positive displacement pumps

$$P\text{, psi} = \frac{\text{H, ft} \times W}{W} \tag{2.13}$$

2.4.4 CALCULATING HORSEPOWER AND EFFICIENCY

When considering work being done, we consider the "rate" at which work is being done. This is called *power* and is labeled as **foot-pounds/second.** At some point in the past, it was determined that the ideal work animal, the horse, could move 550 pounds 1 foot in 1 second. Because large amounts of work are also to be considered, this unit became known as *horsepower.*

When pushing a certain amount of water at a given pressure, the pump performs work. One horsepower equals 33,000 ft-lb/min. The two basic terms for horsepower are

- **hydraulic horsepower, whp**
- **brake horsepower, bhp**

2.4.4.1 HYDRAULIC HORSEPOWER (WHP)

One hydraulic horsepower equals the following:

- 550 ft-lb/sec
- 33,000 ft-lb/min
- 2545 British thermal units per hour (Btu/hr)
- 0.746 kw
- 1.014 metric hp

To calculate the hydraulic horsepower (whp) using flow in gpm and head in feet, use the following formula for centrifugal pumps:

$$\text{whp} = \frac{\text{flow, gpm} \times \text{head, ft} \times \text{specific gravity}}{3960} \tag{2.14}$$

When calculating horsepower for positive displacement pumps, common practice is to use psi for pressure. Then, the hydraulic horsepower becomes

$$\text{whp} = \frac{\text{flow, gpm} \times \text{pressure, psi}}{3960} \qquad (2.15)$$

2.4.4.2 PUMP EFFICIENCY AND BRAKE HORSEPOWER (BHP)

When a motor-pump combination is used (for any purpose), neither the pump nor the motor will be 100% efficient. Simply, not all the power supplied by the motor to the pump (called *brake horsepower,* bhp) will be used to lift the water (*water* or *hydraulic horsepower*)—some of the power is used to overcome friction within the pump. Similarly, not all of the power of the electric current driving the motor (called *motor horsepower,* mhp) will be used to drive the pump—some of the current is used to overcome friction within the motor, and some current is lost in the conversion of electrical energy to mechanical power.

IMPORTANT **Note:** *Depending on size and type, pumps are usually 50–85% efficient, and motors are usually 80–95% efficient. The efficiency of a particular motor or pump is given in the manufacturer's technical manual accompanying the unit.*

A pump's *brake horsepower* (bhp) equals its hydraulic horsepower divided by the pump's efficiency. Thus, the bhp formulas become

$$\text{bhp} = \frac{\text{flow, gpm} \times \text{head, ft} \times \text{specific gravity}}{3960 \times \text{efficiency}} \qquad (2.16)$$

or

$$\text{bhp} = \frac{\text{flow, gpm} \times \text{pressure, psig}}{1714 \times \text{efficiency}} \qquad (2.17)$$

IMPORTANT **Important Point:** *Horsepower requirements vary with flow. Generally, if the flow is greater, the horsepower required to move the water would be greater.*

EXAMPLE 2.9

Problem: Calculate the bhp requirements for a pump handling salt water and having a flow of 600 gpm with 40-psi differential pressure. The specific gravity of salt water at 68°F equals 1.03. The pump efficiency is 85%.

Solution: To use Equation (2.16), convert the pressure differential to total differential head, tdh = 40 × 2.31/1.03 = 90 ft (rounded).

$$\text{bhp} = \frac{600 \times 90 \times 1.03}{3960 \times 0.85}$$

$$= 16.5 \text{ hp (rounded)}$$

Using Equation (2.17)

$$\text{bhp} = \frac{600 \times 40}{1714 \times 0.85}$$

$$= 16.5 \text{ hp (rounded)}$$

When the motor, brake, and motor horsepowers are known and the **efficiency** is unknown, a calculation to determine motor or pump efficiency must be done. Equation (2.18) is used to determine percent efficiency.

$$\textbf{Percent Efficiency} = \frac{\textbf{hp output}}{\textbf{hp input}} \times 100 \qquad (2.18)$$

From Equation (2.18), the specific equations to be used for motor, pump, and overall efficiency are

$$\textbf{Percent Motor Efficiency} = \frac{\textbf{bhp}}{\textbf{mhp}} \times 100 \qquad (2.19)$$

$$\text{Percent Pump Efficiency} \frac{\text{whp}}{\text{bhp}} = \times 100 \qquad (2.20)$$

$$\text{Percent Overall Efficiency} \frac{\text{whp}}{\text{mhp}} = \times 100 \qquad (2.21)$$

EXAMPLE 2.10

Problem: A pump has a water horsepower requirement of 8.5 whp. If the motor supplies the pump with 12 hp, what is the efficiency of the pump?

Solution:

$$\text{Percent pump efficiency} = \frac{\text{whp output}}{\text{bhp supplied}} \times 100$$

$$= \frac{8.5 \text{ whp}}{12 \text{ bhp}} \times 100$$

$$= 0.71 \times 100$$

$$= 71\% \text{ (rounded)}$$

EXAMPLE 2.11

Problem: What is the efficiency if an electric power equivalent to 25-hp is supplied to thev motor and 14-hp of work is accomplished by the pump?

Solution: Calculate the percent of overall efficiency:

$$\text{Percent overall efficiency} = \frac{\text{hp output}}{\text{hp supplied}} \times 100$$

$$= \frac{14 \text{ whp}}{25 \text{ mhp}} \times 100$$

$$= 0.56 \times 100$$
$$= 56\%$$

EXAMPLE 2.12

Problem: Twelve-kw (kilowatts) of power is supplied to the motor. If the brake horsepower is 14-hp, waht is the efficiency of the motor?

Solution. First, convert the kilowatts power to horsepower. Based on the fact that 1 hp = 0.746 kw, the equation becomes

$$\frac{12 \text{ kw}}{0.746 \text{ kw/hp}} = 16.09 \text{ hp}$$

Now, calculate the percent efficiency of the motor:

$$\text{Percent efficiency} = \frac{\text{hp output}}{\text{hp supplied}} \times 100$$

$$= \frac{14 \text{ bhp}}{16.09 \text{ mhp}} \times 100$$

$$= 87\%$$

2.4.5 SPECIFIC SPEED

Specific speed (N_s) refers to an impeller's speed when pumping 1 gpm of liquid at a differential head of 1 ft. Use the following equation for specific speed, where H is at the best efficiency point:

$$N_s = \frac{\text{rpm} \times Q^{0.5}}{H^{0.75}} \qquad (2.22)$$

where

rpm = revolutions per minute
Q = flow (in gpm)
H = head (in ft)

Pump-specific speeds vary between pumps. No absolute rule sets the specific speed for different kinds of centrifugal pumps. However, the following N_s ranges are quite common.

- volute, diffuser, and vertical turbine = 500–5000
- mixed flow = 5000–10,000
- propeller pumps = 9000–15,000

IMPORTANT

Important Note: *The higher the specific speed of a pump, the higher its efficiency.*

2.5 PUMP-CHARACTERISTIC CURVES

The interrelations of pump head, flow, efficiency, and horsepower are known as the *characteristics of the pump*. These are important elements in pump performance, and they are diagrammed graphically on a **performance curve**. The characteristics commonly shown on a pump curve are

- capacity (flow rate)
- total head
- power (brake horsepower)
- efficiency

- speed (Note: speed is only a characteristic if the pump is driven by a variable-speed motor—for our purposes, we will assume that the pump is driven by a constant-speed motor, so the graphs used have only four curves.)

Important Point: *The four pump characteristics that we are concerned with here (capacity, head, power, and efficiency) are related to each other. This is an extremely important point as it is this interrelationship that enables the four pump curves to be plotted on the same graph.*

Experience has shown some important relationships between capacity, head, power, and efficiency. These relationships are

- The capacity (flow rate) of a pump changes as the head against which the pump is working changes.
- Pump capacity also changes as the power supplied to the pump changes.
- Pump capacity changes as efficiency changes.

Consequently, head, power, and efficiency can all be graphed as a function of pump capacity. That is, capacity, Q, designated in gallons per minute, cubic meters per second, is shown along the horizontal (bottom—the X axis) scale of the graph. Head (in pounds per square inch, feet of water, or other pressure designations), power, and efficiency (any one or a combination of them) are shown along the vertical (side—the Y axis) scales of the graph.

Important Point: *Performance curves for centrifugal pumps are different in kind from curves drawn for positive displacement pumps. This is the case because the centrifugal pump is a dynamic device, in that the performance of the pump responds to forces of acceleration and velocity.*

2.5.1 HEAD-CAPACITY (*H-Q*) CURVE

Head-capacity, abbreviated as *H-Q*, is the curve indicating the relationship between total head H, or pressure, against which the pump must

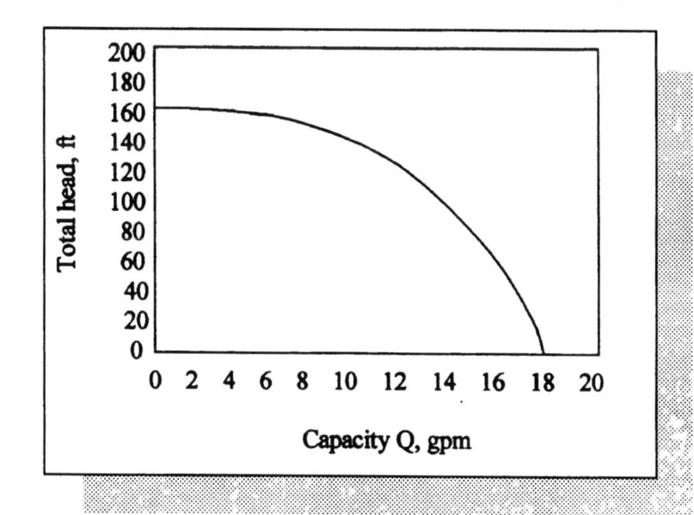

Figure 2.5
H-Q curve.

operate and pump capacity Q. Figure 2.5 shows a typical H-Q curve. The curve indicates what flow rate the pump will produce at any given total head.

The curve for a centrifugal pump may slope to the left or the right, or it may be a flattish curve, depending on the specific speed of the impeller. As capacity increases, the total head that the pump is capable of developing is reduced.

As shown in Figure 2.5, the capacity of the pump decreases as the total head increases (i.e., when the force against which the pump must work increases, the flow rate decreases). The way total head controls the capacity is a characteristic of a particular pump.

IMPORTANT

Important Point: *For pumps, except those having a flattish curve, the highest head occurs at the point where there is no flow through the pump; that is, when the pump is running with the discharge valve closed (i.e., cut-off head; see definition Section 2.2).*

2.5.2 THE POWER-CAPACITY (P-Q) CURVE

The *power-capacity curve*, abbreviated P-Q (Figure 2.6), shows the relationship between power P and capacity Q. In this figure, pump capacity is measured as gallons per minute, and power is measured as brake horsepower.

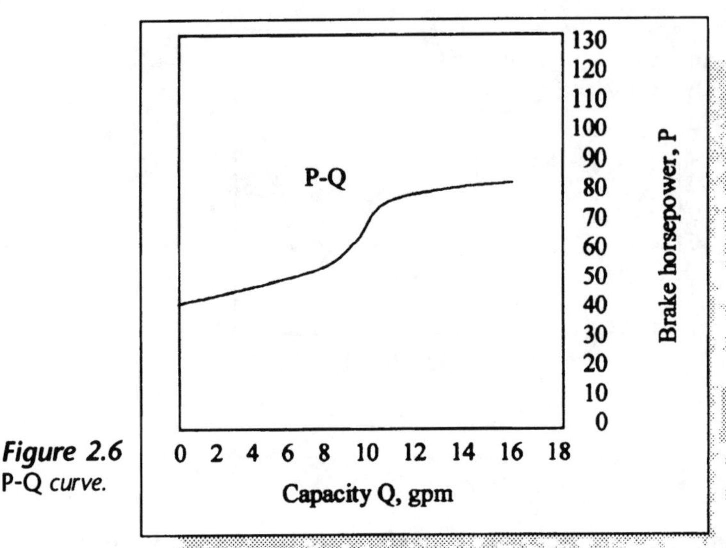

Figure 2.6
P-Q *curve.*

IMPORTANT

Note: *Knowledge of what power the pump requires is valuable for checking the adequacy of an existing pump and motor system.*

2.5.3 THE EFFICIENCY-CAPACITY (*E*-*Q*) CURVE

The *efficiency-capacity curve*, abbreviated *E-Q* (Figure 2.7), shows the relationship between pump efficiency *E* and capacity *Q*. In sizing a pump system, the design engineer attempts to select a pump that will produce the desired flow rate at or near peak pump efficiency.

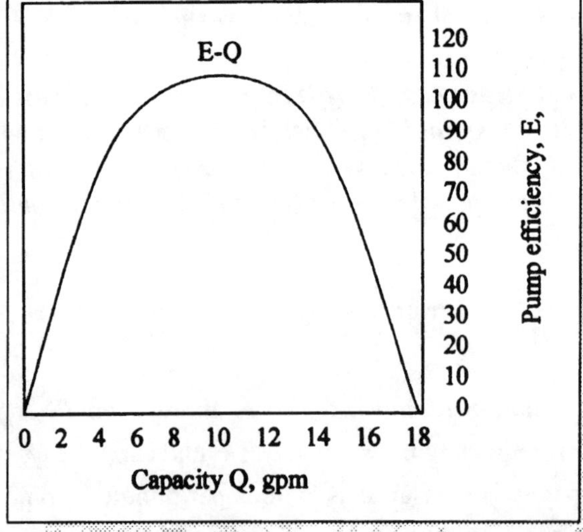

Figure 2.7
E-Q *curve.*

Note: *The more efficient the pump, the less costly it is to operate. (Note: The head-capacity, power-capacity, and efficiency-capacity curves are discussed in detail in Sections 3.5.1–3.5.3).*

2.6 PUMPS IN SERIES AND PARALLEL

Series pump operation is achieved by having one pump discharge into the suction of the next. This arrangement is used primarily to increase the discharge head (i.e., when system heads are too great for one pump to overcome), although a small increase in capacity also results.

Parallel operation is obtained by having two pumps discharging into a common header. Parallel operation is typically employed when head is insufficient, but more flow is needed. Pumps arranged in parallel increase the flow, but the head remains that of one pump working.

Series or parallel operations allow the operator to be flexible enough in pumping capacities and heads to meet requirements of system changes and extensions. With two pumps in parallel, one can be shut down during low demand. This allows the remaining pump to operate close to its optimum efficiency point.

2.7 CONSIDERATIONS FOR PUMPING WASTEWATER[8]

In pumping water, the primary consideration is to ensure that the pumping equipment is operating properly, supply service is readily available, and the pumping equipment is well maintained.

In pumping wastewater, many of the considerations are the same as with pumping water. However, the primary consideration in pumping wastewater is the pump's tendency to clog. Centrifugal pumps for wastewater (i.e., water with large solids) should always be of the single-suction type with non-clog, open impellers. (Note: Double suction pumps are prone to clogging because rags will catch and wrap around the shaft that

[8]Lindeburg, M. R., *Civil Engineering Reference Manual,* 4th ed. San Carlos, CA: Professional Publications, Inc., pp. 4.15–4.16, 1986.

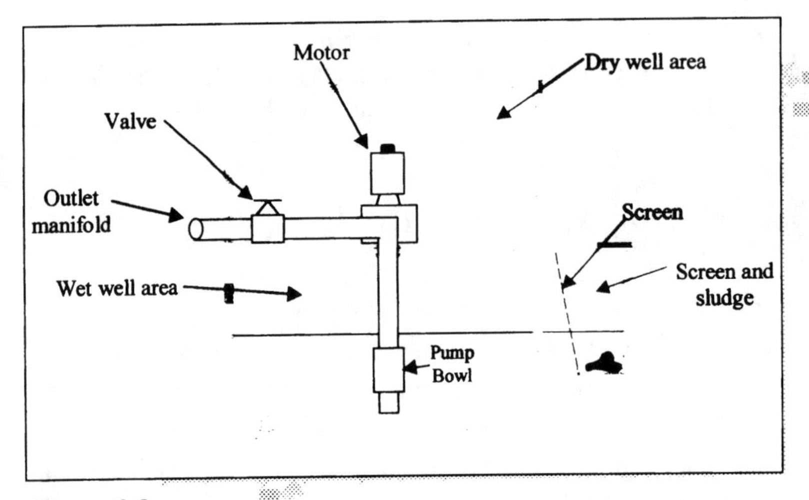

Figure 2.8
Simplified wastewater pump installation.

extends through the impeller eye.) A typical simplified wastewater pump configuration is shown in Figure 2.8. Limiting the number of impeller vanes to two or three, providing for large passageways, and using a bar screen ahead of the pump can further reduce clogging.

The number of pumps used in a wastewater installation is largely dependent on expected demand, pump capacity, and design criteria for backup operation. The number of pumps and their capacities should be able to handle the peak flow with one pump in the set out of service.

2.8 TYPES OF PUMPS USED IN WATER/ WASTEWATER TREATMENT

According to the Hydraulic Institute, all pumps may be classified as kinetic energy pumps or positive displacement pumps.[9] Table 2.2 provides a brief description and application of many types of pumps in these two classes. Basic configurations of many types of pumps are also shown in Figures 2.9 through 2.15.

[9] *Hydraulic Institute Standards for Centrifugal, Rotary and Reciprocation Pumps,* 14th ed., Cleveland, Ohio, p. 3, 1983.

TABLE 2.2. Pump Types and Major Applications in Water/Wastewater.*

Major Classification	Pump Type	Major Pumping Applications
Kinetic	Centrifugal	Raw water/wastewater, secondary sludge return and wasting, settled primary, and thickened sludge, effluent
	Peripheral	Scum, grit, sludge, and raw water/wastewater
	Rotary	Lubricating oils, gas engines, chemical solutions, small flows of water and wastewater
Positive Displacement	Screw	Grit, settled primary and secondary sludges, thickened sludge, raw wastewater
	Diaphragm	Chemical solution
	Plunger	Scum, and primary, secondary, and settled sludges. Chemical solutions.
	Airlift	Secondary sludge circulation and wasting, grit
	Pneumatic ejector	Raw wastewater at small installation (100 to 600 L/min).

*Adapted from Qasim, 1994, pp. 178–179.

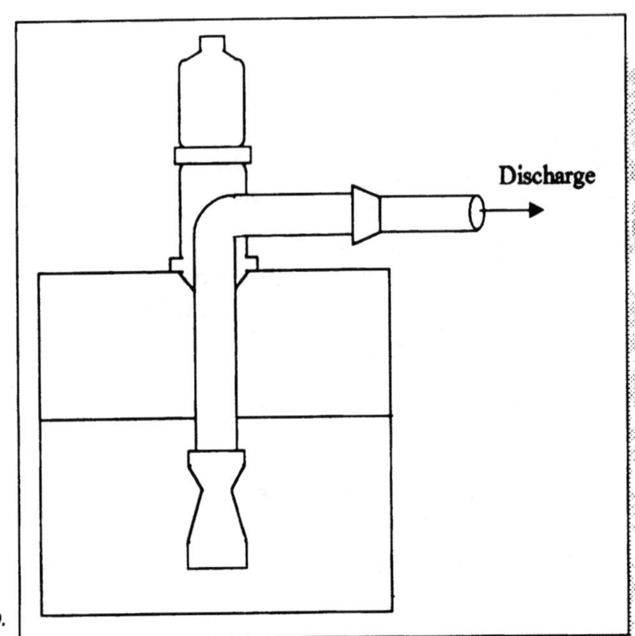

Figure 2.9
Wet-well suspended pump.

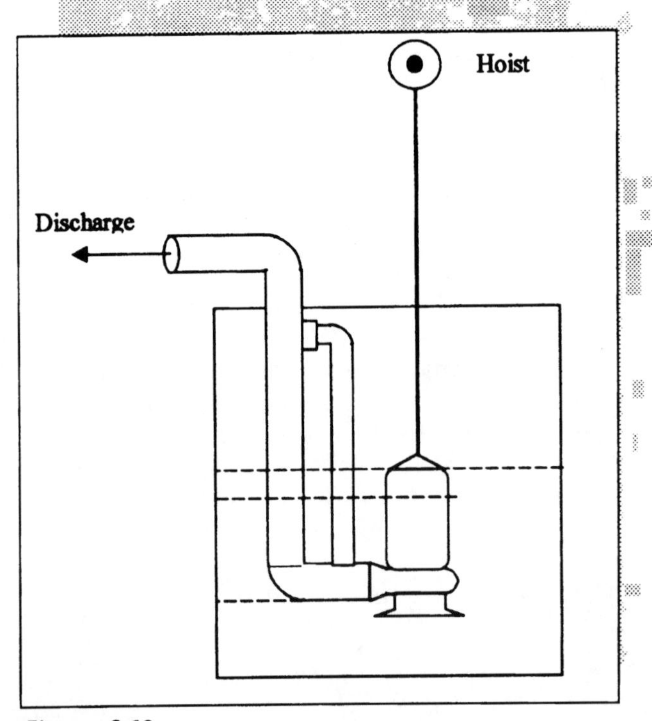

Figure 2.10
Wet-well submersible pump.

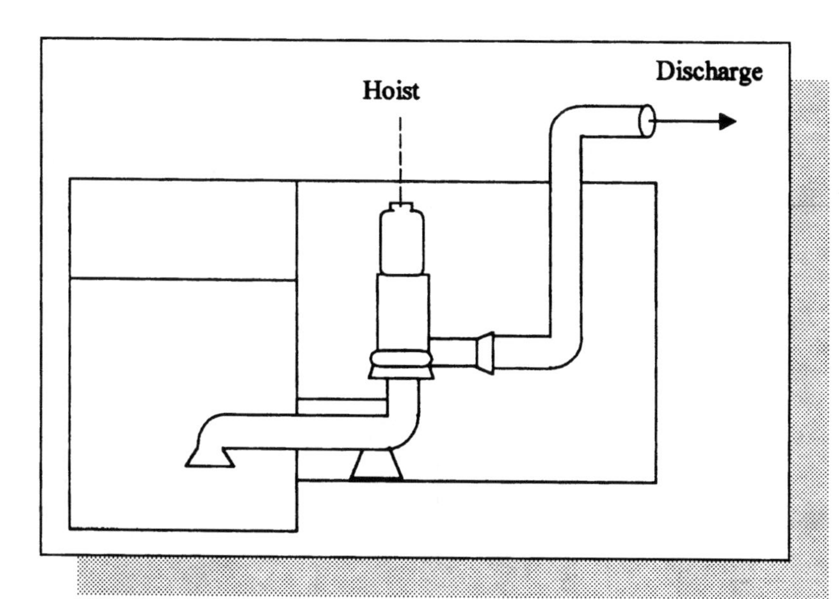

Figure 2.11
Dry-well centrifugal pump.

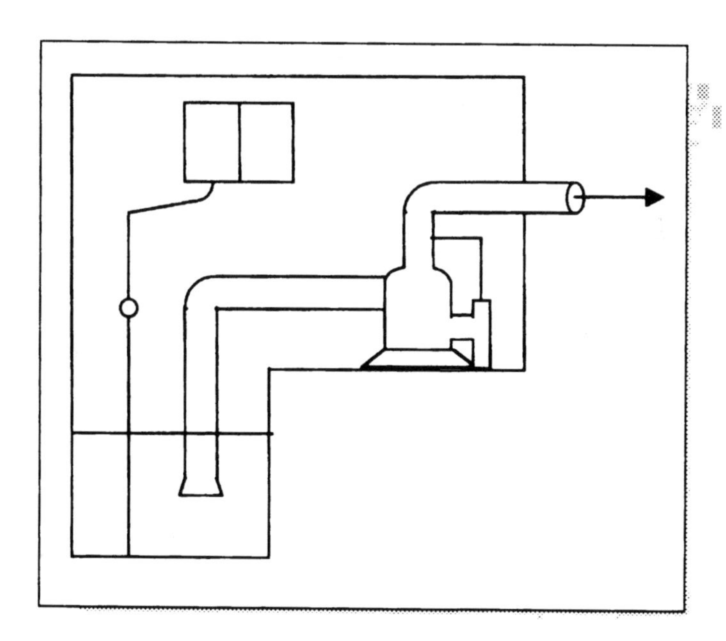

Figure 2.12
Dry-well self-priming pump.

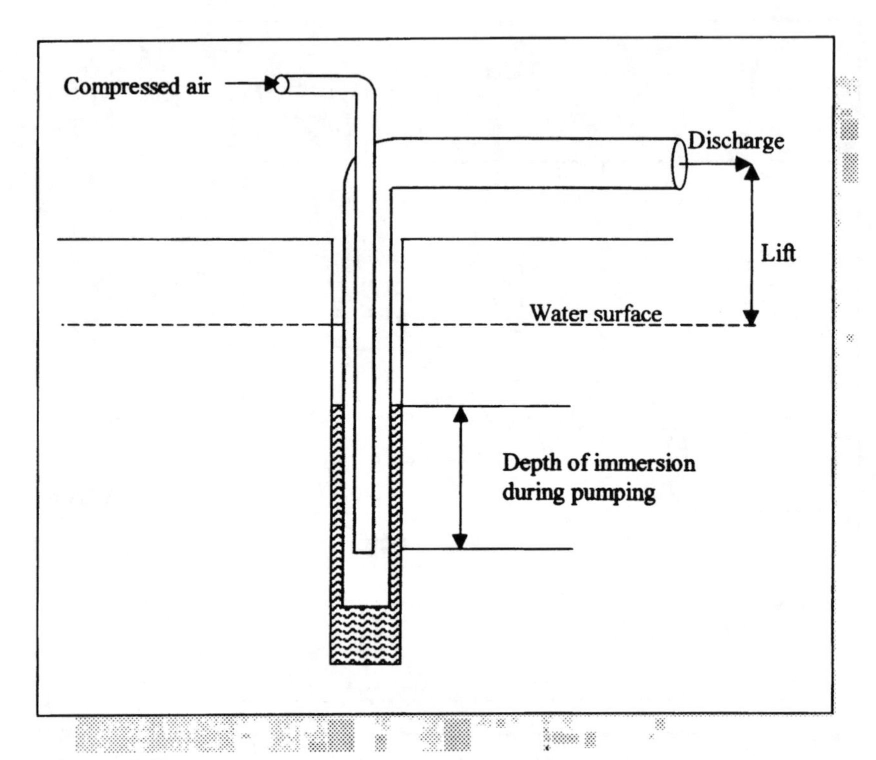

Figure 2.13
Air-lift pump.

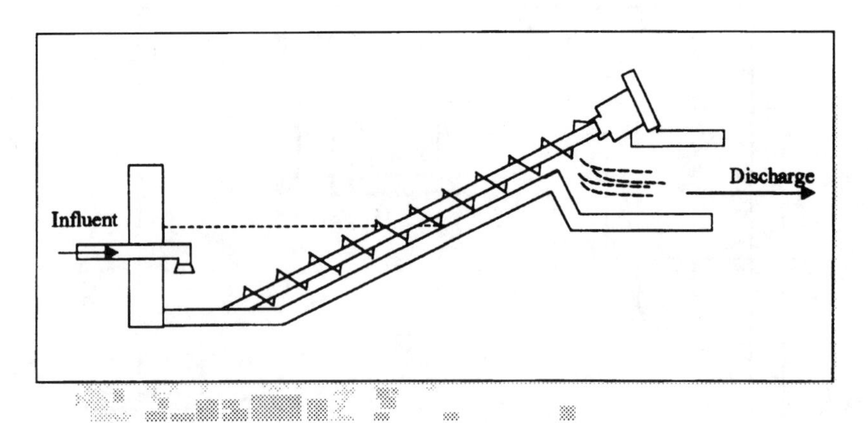

Figure 2.14
Screw pump.

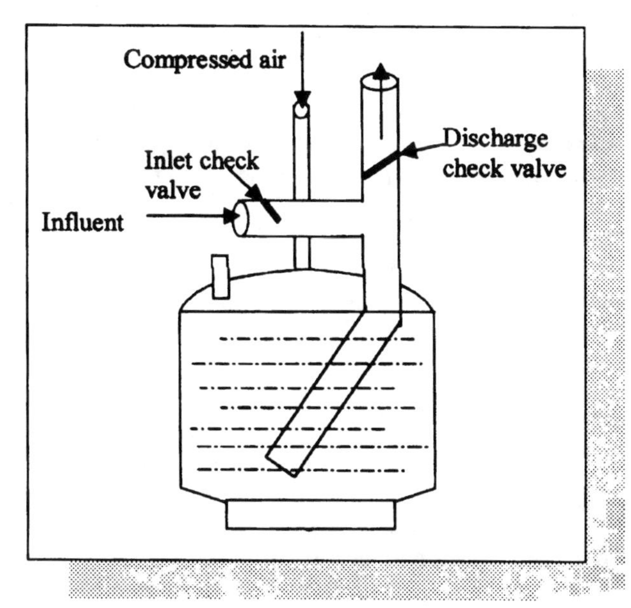

Figure 2.15
Pneumatic ejector.

REFERENCES

Garay, P. N., *Pump Application Book.* Lilburn, GA: The Fairmont Press, 1990.

Hydraulic Institute Standards for Centrifugal, Rotary and Reciprocation Pumps, 14th ed., Cleveland, Ohio, 1983.

Lindeburg, M. R., *Civil Engineering Reference Guide,* 4th ed. San Carlos, CA: Professional Publishers, 1986.

Qasim, S. R., *Wastewater Treatment Plants: Planning, Design, and Operation.* Lancaster, PA: Technomic Publishing Co., Inc., 1994.

Wahren, U., *Practical Introduction to Pumping Technology.* Houston: Gulf Publishing Company, 1997.

The Waterworks Operator Short Course. Blacksburg, VA: Virginia Tech, 1999.

Self-Test

2.1 **Matching Exercise:** Match the definitions listed in part A with the terms listed in part B by placing the correct letter in the blanks.

Part A:

1. The volume or amount of a liquid moving through a channel or pipe:______
2. The total head on the discharge side of the pump: ______
3. The difference between the water pressure and the atmospheric pressure:______
4. The pressure of the atmosphere on a surface:______
5. Indicates roughness of pipe wall:______
6. Equals 30 in. Hg:______
7. The discharge at each point in a pipe or channel is the same as the discharge at any other point:______
8. The pressure exerted on a surface area by the weight of the atmosphere:______
9. Work a pump performs while moving a determined amount of liquid at a given pressure:______
10. On a mountain, air pressure ______ because the blanket is not as thick.
11. A reduction in pressure:______
12. Produces noise and may also cause the pump to vibrate and to lose hydrodynamic efficiency:______
13. Input horsepower delivered to the pump shaft:______
14. Energy supplied by a pump and the energy required to move the liquid to a specified point are equal and no discharge at the desired point occurs:______
15. Height of a column of water in feet:______

16. Speed at least 20% higher or lower than rated speed:______

17. The pressure at any point in a fluid at rest depends on the distance measured vertically to the free surface and the density of the fluid:______

18. The energy required to move a fluid from the supply tank to the discharge point:______

19. The correlation of pump capacity, head, and speed at optimum efficiency:______

20. The vertical distance a liquid can be raised by a given pressure:______

21. The result of dividing the weight of an equal volume of water:______

22. The speed of the fluid moving through a pipe or channel:______

23. Usually expressed in terms of horsepower (hp):______

24. Contains 7.48 gallons:______

25. Using energy to move an object over a distance:______

26. Pressure measured at the pump's discharge:______

Part B:

a. absolute pressure

b. atmospheric pressure

c. cavitation

d. critical speed

e. discharge pressure

f. flow

g. cut off head

h. pressure head

i. static head

j. discharge head

k. horsepower

l. brake hp

m. power

n. specific gravity

o. specific speed

p. full vacuum

q. velocity

r. work

s. decreases

t. 1 cubic foot of water

u. head

v. Stevin's Law

w. gauge pressure

x. The Law of Continuity

y. head loss or friction loss

z. C-factor

2.2 The height of the column of water that will produce 22 psi is ________________.

2.3 What is the pressure at a point 16 feet below the surface of a water storage tank?

2.4 Calculate the bhp requirements for a pump handling salt water and having a flow of 800 gpm with 30-psi differential pressure. The specific gravity of the salt water at 68°F equals 1.03. The pump efficiency is 70%.

2.5 A pump has a water hp requirement of 9.0 whp. If the motor supplies the pump with 10 hp, what is the efficiency of the pump?

Introduction to Centrifugal Pumps

Fire makes things hot and bright and uses them up. Air makes things cool and sneaks in everywhere. Earth makes things solid and sturdy, so they'll last. But water, it tears things down, it falls from the sky and carries off everything it can, carries it off and down to the sea. If the water had its way, the whole world would be smooth, just a big ocean with nothing out of the water's reach. All dead and smooth.[10]

TOPICS

Description
Theory
Types of Pumps
Characteristics and Performance Curves
Advantages/Disadvantages
Water/Wastewater Applications

3.1 INTRODUCTION

The *centrifugal pump* and its modifications (see Chapter 10) are the most widely used type of pumping equipment in the water and wastewater industries. Pumps of this type are capable of moving high volumes of water in a relatively efficient manner. The centrifugal pump is very dependable, has relatively low maintenance requirements, and can be constructed out of a wide variety of materials. The centrifugal pump is available in a wide range of sizes: capacities ranging from a few gpm up to several thousand lb/in^2 [11]. It is considered one of the most dependable systems available for water and wastewater liquid transfer.

The general characteristics of the centrifugal pump are listed in Table 3.1.

[10]Card, O. S., *Seventh Son.* New York: Tor Books, p. 25, 1987.

[11]Cheremisinoff, N. P. and Cheremisinoff, P. N., *Pumps/Compressors/Fans: Handbook.* Lancaster, PA: Technomic Publishing Co., Inc., p. 13, 1989.

Key Terms Used in This Chapter

CENTRIFUGAL PUMP ▶ A pumping mechanism whose rapidly spinning impeller imparts a high velocity to the water that enters, then converts that velocity to pressure upon exit

BASE PLATE ▶ The foundation under a pump. It usually extends far enough to support the drive shaft. This base plate may also be referred to as the pump frame.

BEARINGS ▶ Devices used to reduce friction and to allow the shaft to rotate easily. Bearings may be ball, roller, or sleeve.

- **Inboard Bearing** In a single suction pump, it is the bearing farthest from the coupling.
- **Outboard Bearing** In a single suction pump, it is the bearing located nearest the coupling.
- **Thrust Bearing** In a single suction pump, it is the bearing located nearest the coupling.
- **Radial Bearing** In a single suction pump, it is the bearing located farthest from the coupling

Note: In most cases, where pump and motor are constructed on a common shaft (no coupling), the bearings will be part of the motor assembly.

CASING ▶ The housing surrounding the rotating element of the pump. In the majority of centrifugal pumps, this casing can also be called the **volute.**

- **Split Casing** A pump casing that is manufactured in two pieces that are fastened together by means of bolts. Split-casing pumps may be vertically (perpendicular to the shaft direction) split or horizontally (parallel to the shaft direction) split.

COUPLING ▶ A device to join the pump shaft to the motor shaft.

- **Close Coupled** Pump and motor are constructed on a common shaft.

EXTENDED SHAFT ▶ Pump is constructed on one shaft and must be connected to the motor by a coupling.

FRAME	The housing that supports the pump bearing assemblies. In an end suction pump, it may also be the support for the pump casing and the rotating element.
IMPELLER	The rotationg element in the pump, which actually transfers the energy from the drive unit to the liquid. Depending on the pump application, the impeller may be open, semi-open, or closed. It may also be single or double suction.
IMPELLER EYE	The center of the impeller; the area that is subject to lower pressures due to the rapid movement of the water to the outer edge of the casing
PRIME	Filling the casing and impeller with water. If this area is not completely full of water, the centrifugal pump will not pump efficiently.
SEALS	Devices used to stop the leakage of air into the inside of the casing around the shaft
Packing	Material that is placed around the pump shaft to seal the shaft opening in the casing and prevent air leakage into the casing
STUFFING BOX	The assembly located around the shaft at the rear of the casing. It holds the packing and lantern ring.
LANTERN RING	Also known as the seal cage, it is positioned between the rings of packing in the suffing box to allow the introduction of a lubricant (water, oil, or grease) onto the surface of the shaft to reduce the friction between the packing and the rotating shaft.
GLAND	Also known as the packing gland, it is a metal assembly that is designed to appy even pressure to the packing to compress it tightly around the shaft.
MECHANICAL SEAL	A device consisting of a stationary element, a rotating element, and a spring to supply force

MECHANICAL SEAL ***(continued)***	to hold the two elements together. Mechanical seals may be either single or double units.
SHAFT	The rigid steel rod that transmits the energy from the motor to the pump impeller. Shafts may be either vertical or horizontal.
SHAFT SLEEVE	A piece of metal tubing placed over the shaft to protect the shaft as it passes through the packing or seal area. In some cases, the sleeve may also help to position the impeller on the shaft.
SHUT-OFF HEAD	The head or pressure at which the centrifugal pump will stop discharging. It is also the pressure developed by the pump when it is operated against a closed discharge valve. This is also known as a cut-off head.
SHROUD	The metal plate that is used to either support the impeller vanes (open or semi-open impeller) or to enclose the vanes of the impeller (closed impeller)
SLINGER	A device to prevent pumped liquids from traveling along the shaft and entering the bearing assembly. Also called a deflector or slinger ring.
WEARING RINGS	Devices that are installed on stationary or moving parts within the pump casing to protect the casing and/or the impeller from wear due to the movement of water through points of small clearances
Impeller Ring	A wearing ring installed directly on the impeller
Casing Ring	A wearing ring installed in the casing of the pump; also known as the suction head ring
Stuffing Box Cover Ring	A wearing ring installed at the rear of the impeller in an end suction pump to maintain the impeller clearances and to prevent casing wear

TABLE 3.1. Characteristics of Centrifugal Pumps.*			
Characteristic	**Description**	**Characteristic**	**Description**
Flow rate	High	Self-priming	No
Pressure rise per stage	Low	Outlet stream	Steady
Constant variable over operating range	Pressure rise	Works with high-viscosity fluids	No

*Adapted from Lindeburg, 1986, p. 4–2.

This chapter will (1) describe the centrifugal pump, (2) provide a brief discussion of pump theory, (3) describe the types of centrifugal pumps, (4) discuss pump characteristics, (5) describe the advantages and disadvantages of the centrifugal pump, and (6) list centrifugal pump applications.

(Note: Chapters 3 through 11 specifically address centrifugal pumps because the centrifugal pump is the most widely used pump in water supply and wastewater treatment. The positive displacement pump is discussed in Chapter 12.)

3.2 DESCRIPTION

The centrifugal pump consists of a rotating element (impeller) sealed in a casing (volute). The rotating element is connected to a drive unit or prime mover (motor/engine) that supplies the energy to spin the rotating element. As the impeller spins inside the volute casing, an area of low pressure is created in the center of the impeller. This pressure allows the atmospheric pressure on the water in the supply tank to force the water up to the impeller. [Note: The term "water" includes both freshwater (potable) and/or wastewater, unless otherwise specified.] Because the pump will not operate if there is no low-pressure zone created at the center of the impeller, it is important that the casing be sealed to prevent air from entering the casing. To ensure the casing is airtight, the pump includes some type of seal (mechanical or conventional packing) assem-

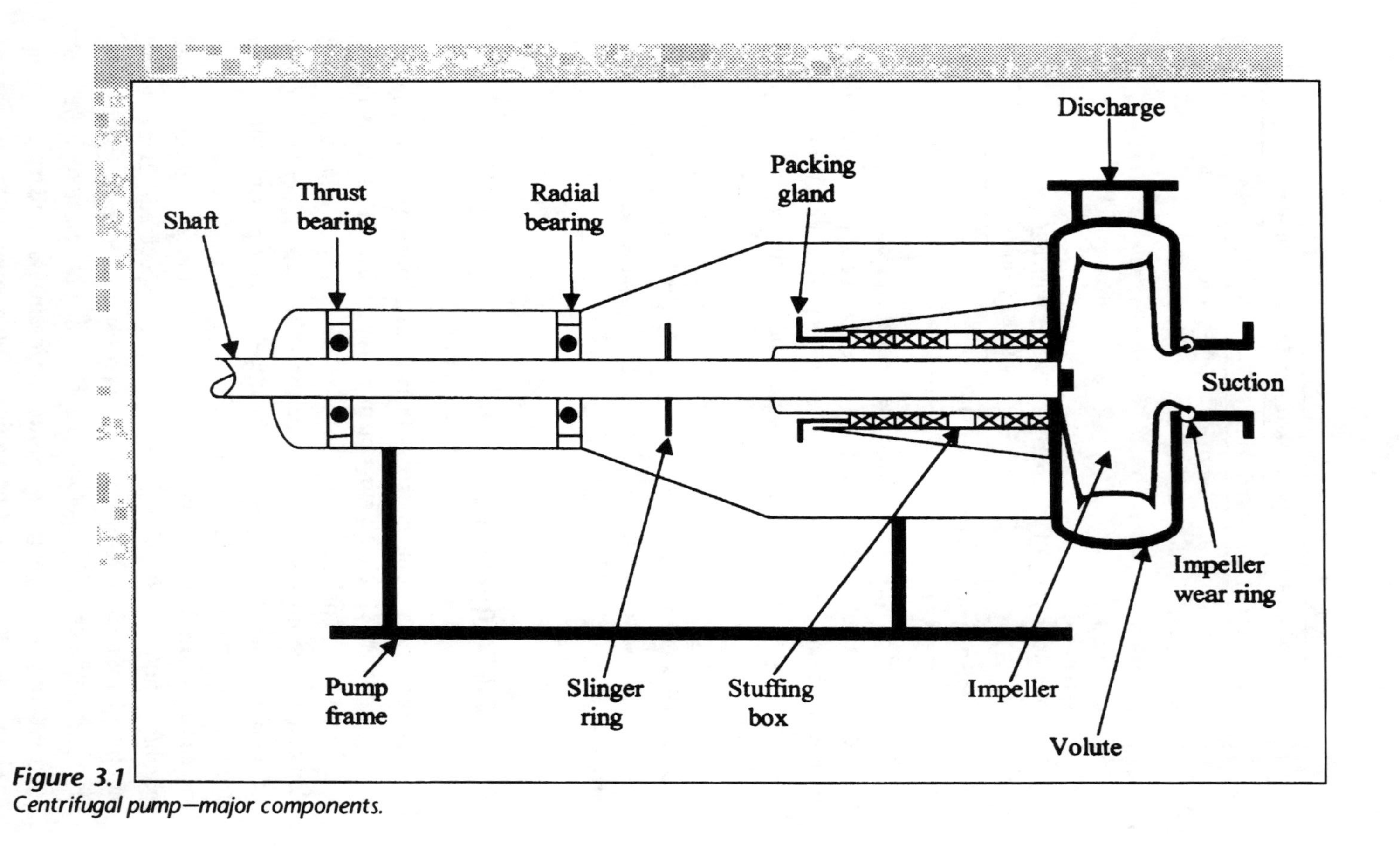

Figure 3.1
Centrifugal pump—major components.

bly at the point where the shaft enters the casing. This seal also includes some type of lubrication (water, grease, or oil) to prevent excessive wear.

When the water enters the casing, the spinning action of the impeller transfers energy to the water. This energy is transferred to the water in the form of increased speed or velocity. The water is thrown outward by the impeller into the volute casing where the design of the casing (Section 4.2.1) allows the velocity of the water to be reduced, which, in turn, converts the velocity energy (velocity head) to pressure energy (pressure head). The process by which this change occurs is described in Section 3.3. The water then travels out of the pump through the pump discharge. The major components of the centrifugal pump are shown in Figure 3.1.

3.3 THEORY

The volute-cased centrifugal pump provides the pumping action necessary (i.e., converts velocity energy to pressure energy) to transfer water from one point to another (see Figure 3.2). The rotation of a series of vanes in an impeller creates pressure. The motion of the impeller forms a partial vacuum at the suction end of the impeller. Outside forces, such as the atmospheric pressure or the weight of a column of liquids, push water into the impeller eye and out to the periphery of the impeller. From there, the rotation of the high-speed impeller throws the water into the pump

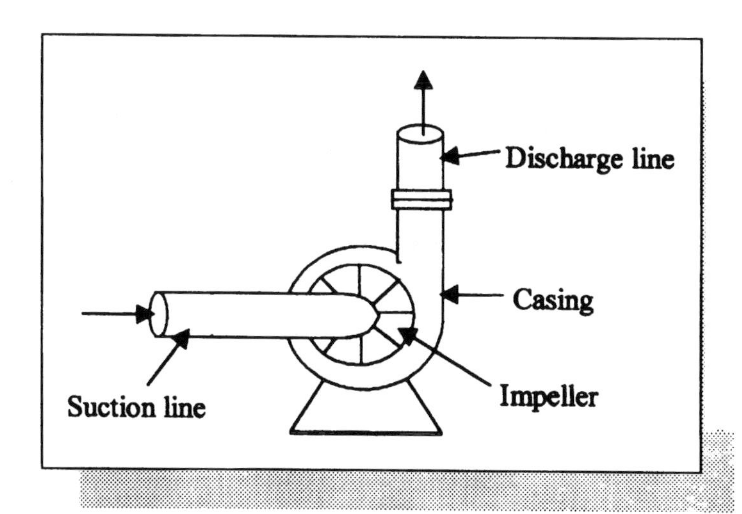

Figure 3.2
A centrifugal pump.

casing. As a given volume of water moves from one cross-sectional area to another within the casing, the velocity or speed of the liquid changes proportionately.

The volute casing has a cross-sectional area that is extremely small at the point in the case that is farthest from the discharge (see Figure 3.3). This area increases continuously to the discharge. As this area increases, the velocity of the water passing through it decreases as it moves around the volute casing to the discharge point.

As the velocity of the liquid decreases, the velocity head decreases, and the energy is converted to pressure head. There is a direct relationship between the velocity of the liquid and the pressure it exerts. Therefore, as the velocity of the liquid decreases, the excess energy is converted to additional pressure (pressure head). This pressure head supplies the energy to move the liquid through the discharge piping.

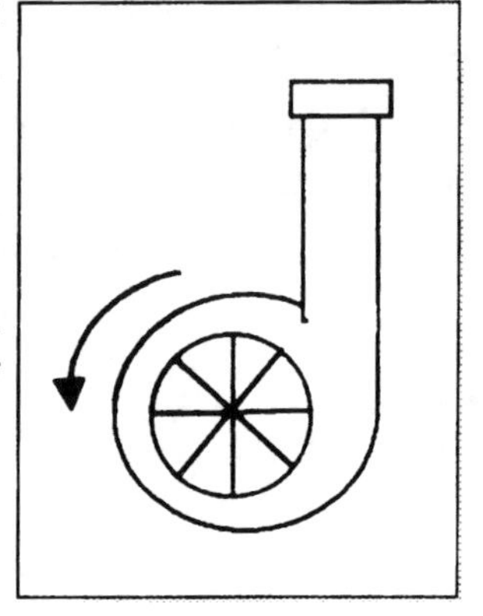

Figure 3.3
Centrifugal pump volute casing.

IMPORTANT ***Important Point:*** *A centrifugal pump will, in theory, develop the same head regardless of the fluid pumped. However, the pressure generated differs (i.e., because of specific gravity differences between various liquids).*

3.4 TYPES OF PUMPS

Centrifugal pumps can be classified into three general categories according to the way the impeller imparts energy to the fluid. Each of these categories has a range of specific speeds and appropriate applications.

The three main categories of centrifugal pumps:

- axial flow impellers
- mixed flow impellers
- radial flow impellers

Any of these pumps can have one or several impellers, which may be

- open
- closed
- semi-open
- single suction
- double suction

3.4.1 RADIAL FLOW IMPELLER PUMPS

Most centrifugal pumps are of radial flow. *Radial flow impellers* impart energy primarily by centrifugal force. Water enters the impeller at the hub and flows radially to the periphery (outside of the casing—see Figure 3.4). Flow leaves the impeller at a 90-degree angle from the direction it enters the pump. Single suction impellers have a specific speed less than 5000. Double suction impellers have a specific speed less than 6000.

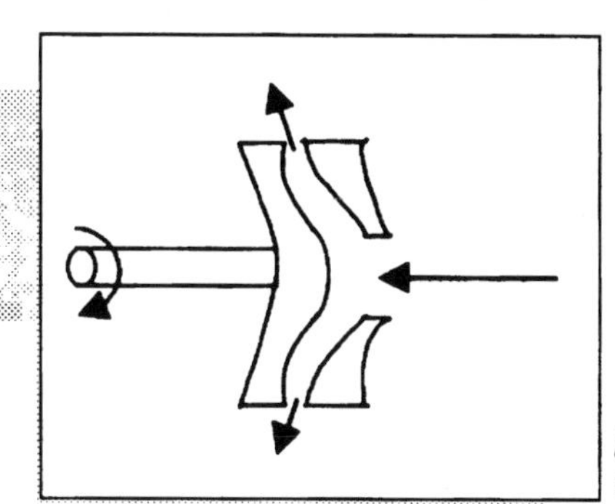

Figure 3.4
Centrifugal (radial) flow pump.

Types of radial flow impeller pumps:

- end suction pumps (see Figure 3.5)
- in-line pumps
- vertical volute pumps (cantilever)
- axially (horizontally) split pumps
- multistage centrifugal pumps
- vertical turbine pumps

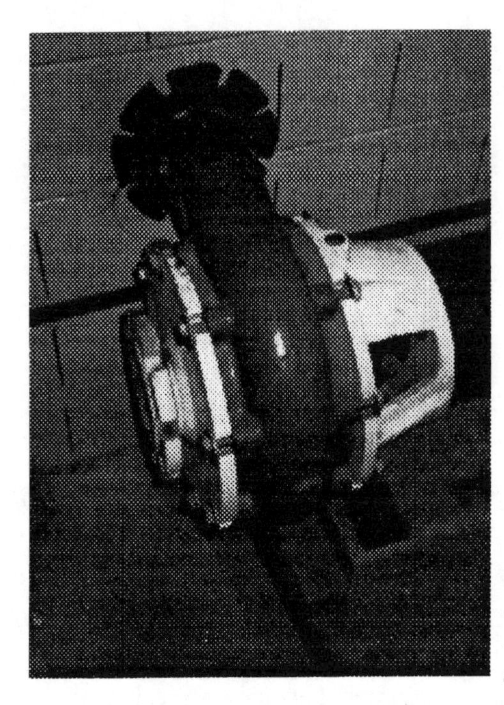

Figure 3.5 *Horizontal end suction centrifugal pump.*

IMPORTANT **Note:** *The high service pump at a potable water treatment plant, which lifts water from the plant to elevated storage, is usually a radial pump.*

3.4.2 MIXED FLOW IMPELLER PUMPS

Mixed flow impellers impart energy partially by centrifugal force and partially by axial force, because the vanes act partially as an axial compressor. This type of pump has a single inlet impeller with the flow entering axially and discharging in an axial and radial direction (see Figure 3.6). Specific speeds of mixed flow pumps range from 4200 to 9000.

IMPORTANT **Note:** *Mixed flow impeller pumps are suitable for pumping untreated wastewater and stormwater. They operate at higher speeds than the radial flow impeller pumps, are usually of lighter construction, and, where applicable, cost less than corresponding non-clog pumps. Impellers may be either open or enclosed, but enclosed is preferred.*[12]

[12]Metcalf & Eddy, *Wastewater Engineering: Collection and Pumping of Wastewater.* Tchobanoglous, G. (ed.). New York: McGraw-Hill Book Company, p. 281, 1981.

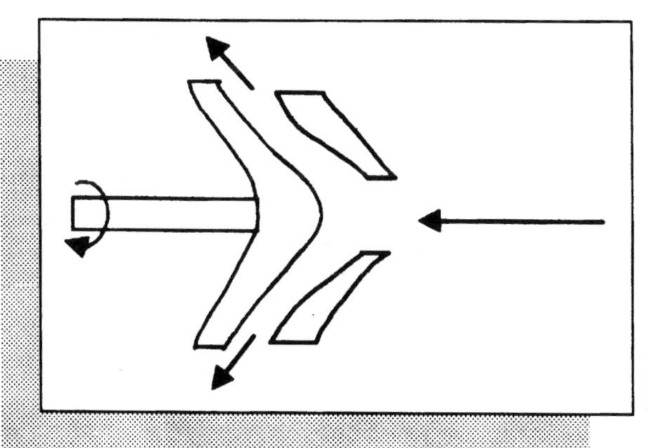

Figure 3.6
Mixed flow pump.

3.4.3 AXIAL FLOW IMPELLER PUMPS (PROPELLER PUMP)

Axial flow impellers impart energy to the water by acting as axial flow compressors (see Figure 3.7). The axial flow pump has a single inlet impeller with flow entering and exiting along the axis of rotation (along the pump drive shaft). Specific speed is greater than 9000. The pumps are used in low-head, large-capacity applications, such as

- municipal water supplies
- irrigation
- drainage and flood control
- cooling water ponds
- backwashing
- low service applications (e.g., they carry water from the source to the treatment plant)

IMPORTANT

Important Point: *Radial flow and mixed flow centrifugal pumps can be designed for either single or double suction operation. In a* ***single suction pump****, water enters only one side of the impeller. In a* ***double suction pump****, water enters both sides of the impeller. Thus, for an impeller with a given specific speed, a greater flow rate can be expected from a double suction pump.*

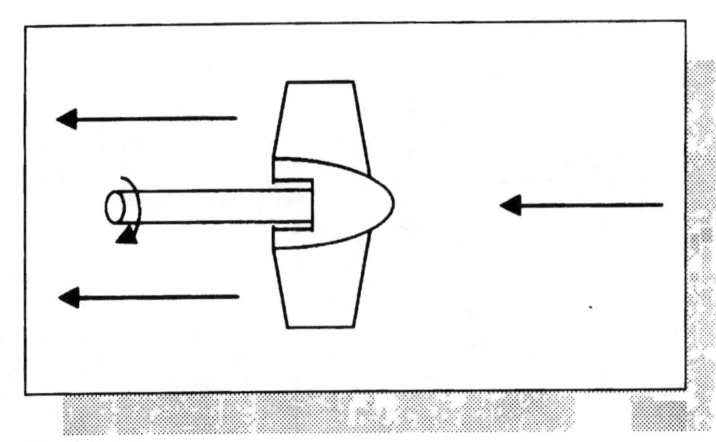

Figure 3.7
Axial flow pump.

3.5 CHARACTERISTICS AND PERFORMANCE CURVES

Recall that Section 2.5 provided a general discussion of pump characteristic curves. The following sections specifically discuss pump characteristics and performance curves directly related to the centrifugal pump.

The centrifugal pump operates on the principle of an energy transfer and, therefore, has certain definite characteristics that make it unique.

Many manufacturers produce pumps of similar size and design, but they vary somewhat because of the design modifications made by each manufacturer. Operating characteristics for various types of centrifugal pumps are reported in Table 3.2.

The type and size of the impeller limit the amount of energy that can be transferred to the water, the characteristics of the material being pumped, and the total head of the system through which the liquid is moving. A series of performance graphs or curves best describes the relationship between these factors. The sections that follow discuss centrifugal pump performance curves in greater detail.

TABLE 3.2. Operating Characteristics for Centrifugal Pumps.*

Characteristics	Radial Flow	Mixed Flow	Axial Flow
Flow	Even	Even	Even
Effect of head on:			
Capacity	Decrease	Decrease	Decrease
Power required	Decrease	Small decrease to large increase	Large increase
Effect of decreasing head on:			
Capacity	Increase	Increase	Increase
Power required	Increase	Slight increase to decrease	Decrease
Effect of closing discharge valve on:			
Pressure	Up to 30% increase	Considerable increase	Large increase
Power required	Decrease 50%–60%	10% decrease, 80% increase	Increase 80–150%

*Adapted from Metcalf & Eddy, 1981, p. 291.

IMPORTANT

Note: *Garay[13] points out that performance curves for centrifugal pumps are different in kind from curves drawn for positive displacement pumps. This is the case because the centrifugal pump is a dynamic device in that the performance of the pump responds to forces of acceleration and velocity. Note that every basic performance curve is based on a particular speed and a specific impeller diameter, impeller width, and fluid viscosity (usually taken as the viscosity of water). While impeller diameter and speed can usually be manipulated within the design of a specific casing, the width of the impeller cannot be changed significantly without selecting a different casing.*

3.5.1 HEAD-CAPACITY CURVE

As might be expected, the capacity of a centrifugal pump is directly related to the *total head* of the system. If the total head on the system is

[13]Garay, P. N., *Pump Application Desk Book.* Lilburn, GA: The Fairmont Press, Inc., p. 143, 1990.

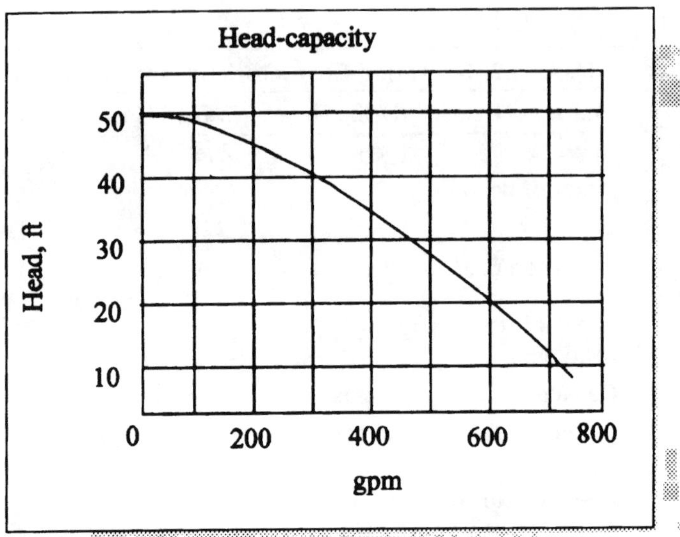

Figure 3.8
Head-capacity curve.

increased, the volume of the discharge will be reduced proportionately. Figure 3.8 illustrates a typical head-capacity curve (commonly abbreviated *H-Q*, as pointed out in Section 2.5.1). While this curve may change with respect to total head and pump capacity based upon the size of the pump, pump speed, and impeller size and/or type, the basic form of the curve will remain the same. As the head of the system increases, the capacity of the pump will decrease proportionately until the discharge stops. The head at which the discharge no longer occurs is known as the *cut-off head.*

As discussed earlier, the total head includes a certain amount of energy to overcome the friction of the system. This friction head can be greatly affected by the size and configuration of the piping and the condition of the system's valving. If the control valves on the system are closed partially, the friction head can increase dramatically. When this happens, the total head increases, and the capacity or volume discharged by the pump decreases. In many cases, this method is employed to reduce the discharge of a centrifugal pump. It should be remembered, however, that this does increase the load on the pump and drive system, causing additional energy requirements and additional wear.

The total closure of the discharge control valve increases the friction head to the point where all the energy supplied by the pump is consumed in the friction head and is not converted to pressure head. Therefore, the pump exceeds its cut-off head, and the pump discharge is reduced to zero. Again, it is important to note that, although the operation of a centrifugal pump against a closed discharge may not be as hazardous as with other types of pumps, it should be avoided due to the excessive load placed on the drive unit and pump. There have also been documented cases where the pump produced pressures higher than the pump discharge piping could withstand. In these cases, the discharge piping was severely damaged by the operation of the pump against a closed or plugged discharge.

3.5.2 EFFICIENCY CURVE

IMPORTANT **Important Point:** Efficiency *represents the percentage of useful water horsepower developed by the horsepower required to drive the pump.*

Every centrifugal pump will operate with varying degrees of efficiency over its entire capacity and head ranges. The important factor in selecting a centrifugal pump is to select a unit that will perform near its maximum efficiency in the expected application. Figure 3.9 illustrates a typical effi-

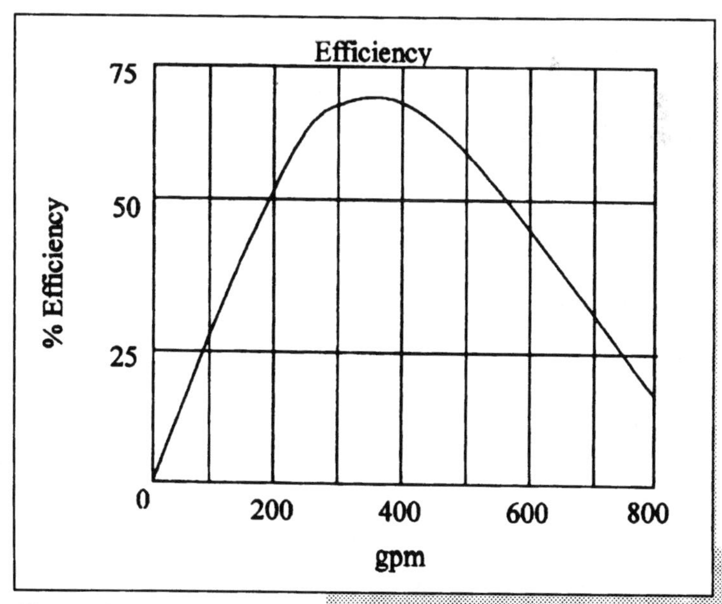

Figure 3.9 Efficiency curve.

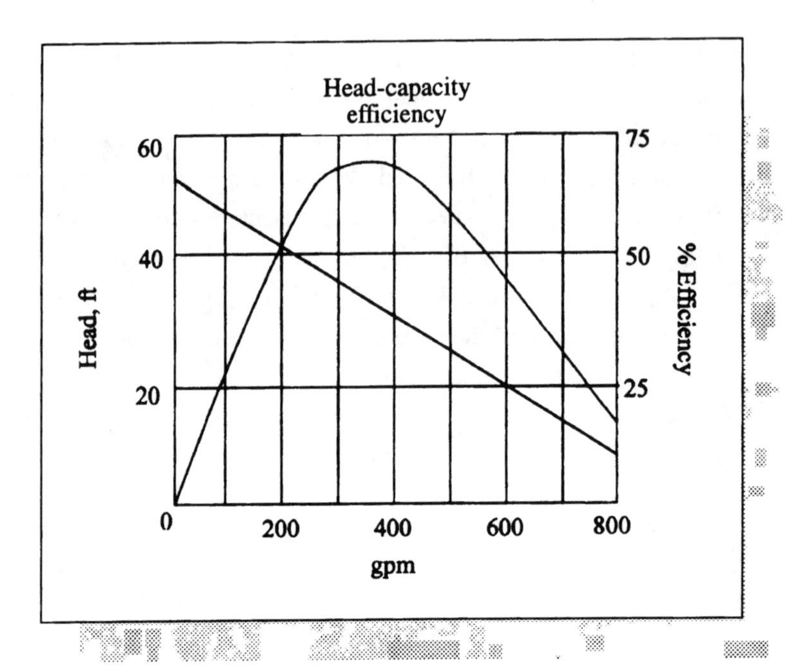

Figure 3.10
Head capacity/efficiency curve.

ciency curve for a centrifugal pump (commonly abbreviated *E-Q* curve, as pointed out in Section 2.5.3). This particular efficiency curve is specific to one pump, with a specified specific speed, impeller size and type and inlet and discharge size. If any of these factors are changed, the efficiency curve for the pump will also change.

For ease of comparison of the head capacity and efficiency curves for a particular pump, it is common practice to print both curves on a single sheet of graph paper as shown in Figure 3.10.

Moreover, it is common practice to use this same procedure to illustrate the head capacity/efficiency curves for a series of pumps that use the same volute casing size and inlet and discharge size, but may have the capability to operate at different speeds or with different sized impellers. In this instance, the head capacity of each pump configuration is shown on the graph with efficiency being shown as zones or regions. Figure 3.11 illustates the combined curves for a single model pump that has the capability to operate with several different size impellers and at several different speeds.

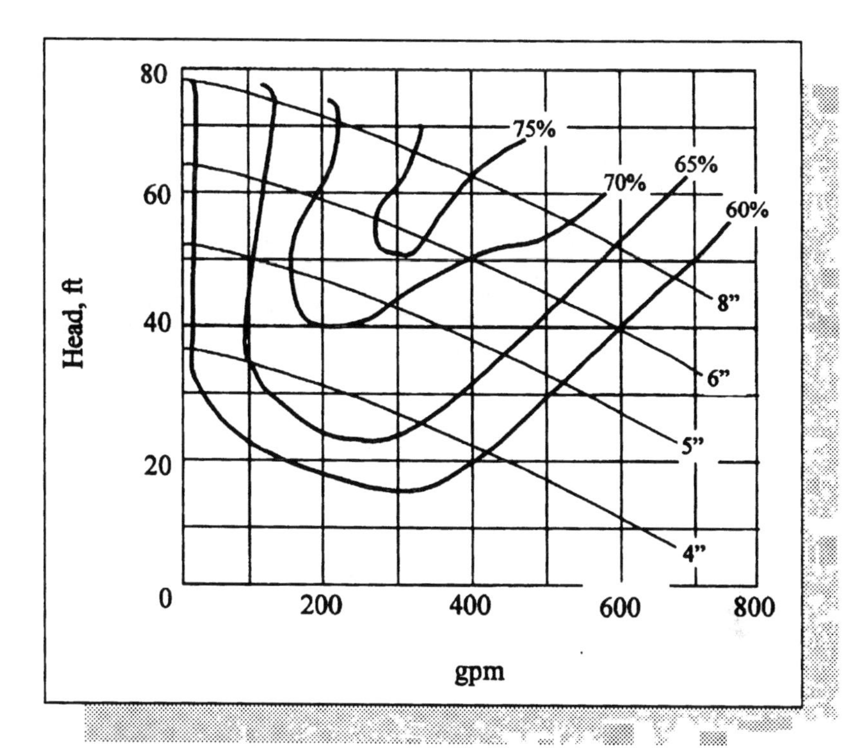

Figure 3.11
Head capacity/efficiency curve.

3.5.3 BRAKE HORSEPOWER (BHP) CURVES

In addition to the head capacity and efficiency curves, most pump literature includes a graph showing the amount of energy in horsepower that must be supplied to the pump to obtain the performance shown in the head capacity curve. To afford easy use of this information, the *brake horsepower curve* is usually incorporated into the previous two curves on a single chart as shown in Figure 3.12.

As was the case for the efficiency curve, this can also be shown for all the pumps within a given series on a single combined chart (commonly abbreviated *P-Q*, power-capacity curve, as discussed in Section 2.5.2). This chart will normally show the required brake horsepower as a series of lines on the chart (see Figure 3.13).

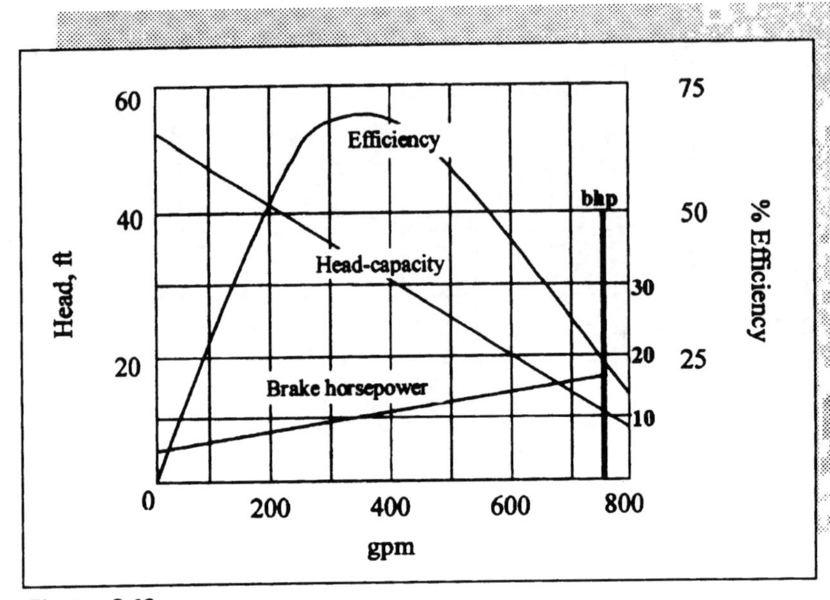

Figure 3.12
Head capacity/efficiency/brake horsepower curve.

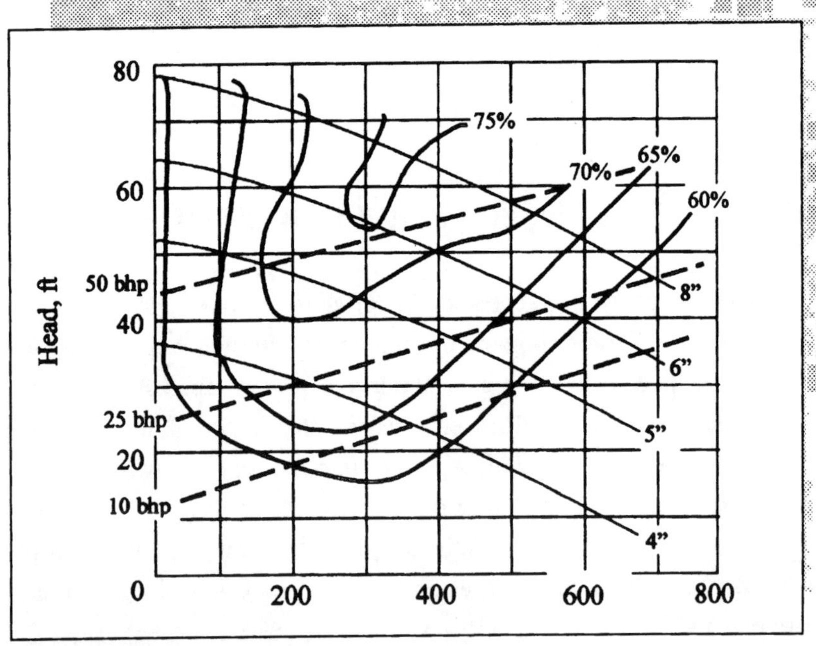

Figure 3.13
Head capacity/efficiency/brake horsepower curve.

3.6 ADVANTAGES/DISADVANTAGES[14]

The centrifugal pump has many advantages that make it one of the most widely used types of pumps. Although it is one of the most widely used pumps, it does have a few disadvantages. Both the advantages and disadvantages are discussed in the following sections.

3.6.1 ADVANTAGES OF THE CENTRIFUGAL PUMP[14]

The advantages of the centrifugal pump stem from its construction, operation, maintenance requirements, wide tolerances for moving parts, self-limitation of pressure, adaptability, space requirements, and its rotary rather than reciprocating motion.

- **Construction:** The centrifugal pump consists of a single rotating element and a simple casing that can be constructed using a wide assortment of materials. If the material to be pumped is highly corrosive, the pump parts that contact the liquid can be constructed of lead or other material that is not likely to corrode. If the material being pumped is highly abrasive (such as grit or ash from an incinerator), the internal parts can be made of abrasion-resistant material or coated with a protective material. Moreover, the simple design of a centrifugal pump allows the pump to be constructed in a wide variety of sizes and configurations. Compared to the centrifugal pump, no other currently available pump has the range of capacities or applications.
- **Operation:** The operation of a centrifugal pump is both simple and relatively quiet. The average operator with a minimum amount of training can operate pumping facilities that use centrifugal-type pumps. Moreover, the pump can withstand a great deal of improper operation without major damage.
- **Maintenance:** Routine preventive maintenance requirements for the centrifugal-type pump are not as demanding as those associated with

[14]Adapted from notes taken while attending the *Basic Maintenance Training Course*—Onondaga County Department of Drainage and Sanitation. North Syracuse, New York, 1984.

some of the other pumping systems. While there is a requirement to perform a certain amount of preventive maintenance, the skills required to perform this maintenance are normally considered less complex than those required for other pumping systems.

- **Wide Tolerance for Moving Parts:** The design of the centrifugal pump does not require that all moving parts be constructed to very close tolerances. Therefore, the amount of wear on these moving parts is reduced, and the operating life of the equipment is extended.
- **Self-Limitation of Pressure:** Due to the nature of the pumping action, the centrifugal pump will not exceed a predetermined maximum pressure. Therefore, if the discharge valve is suddenly closed, the pump cannot generate additional pressure that might result in damage to the system or could potentially result in a hazardous working condition. The power supplied to the impeller will only generate a specified amount of pressure (head). If a major portion of this pressure or head is consumed in overcoming friction or is lost as heat energy, the pump will have a decreased capacity.
- **Adaptable to High-Speed Drive Systems (Electric Motors):** The centrifugal pump allows the use of high-speed, high-efficiency drive systems. In situations where the pump is selected to match a specific operating condition that remains relatively constant, the pump drive unit can be used without the need for expensive speed reducers.
- **Small Space Requirements:** For the majority of pumping capacities, the amount of space required for installation of the centrifugal-type pump is much less than that of any other type of pump.
- **Rotary Rather than Reciprocating Motion:** Because the centrifugal pump has fewer moving parts, space and maintenance requirements are significantly reduced.

3.6.2 DISADVANTAGES OF THE CENTRIFUGAL PUMP

Although the centrifugal pump is the most widely used type of pump in water/wastewater applications, and in most other general industry-wide applications, it does have a few disadvantages. Disadvantages include the following: it is not a self-priming pump, air leaks on the suction side affect pump performance, high efficiency can only be maintained over a narrow range, if stopped without closing the discharge line, the pump may run backwards, and pump speed cannot usually be limited without the use of additional equipment.

- **Not Self-Priming:** Although the centrifugal pump can be installed in a manner that will make the pump self-priming, it is not truly capable of drawing liquid to the pump impeller unless the pump casing and impeller are filled with water. [Note: If there exists a suction head (positive pressure on the suction side of the pump), the unit will always remain full whether on or off, but with a suction lift, water tends to run back out of the pump and down the suction line when the pump stops.] The bottom line: If for any reason the water in the casing and impeller drains out, the pump would cease pumping until this area is refilled.

IMPORTANT

Important Point: *The previous point is important primarily because many people hold the misconception that a centrifugal pump "sucks" water from its source and that it is this "sucking" action that conveys the liquid along its distribution network. Nothing could be further from the truth. The fact that a centrifugal pump must be filled with water (primed) before it can perform its pumping action points out that the pump actually "forces" the water to move, instead of "sucking" the water to move it.*

As a result of its need to be primed, it is normally necessary to start a centrifugal pump with the discharge valve closed. The valve is then gradually opened to its proper operating level. Starting the pump against a closed discharge valve is not hazardous provided the valve is not left closed for extended periods.

IMPORTANT

Important Point: *While it is normally the procedure to leave the valve closed on the startup of a centrifugal pump, this should **never** be done on a positive-displacement pump.*

- **Suction Side Air Leaks:** Air leaks on the suction side of the centrifugal pump can cause reduced pumping capacity in several ways. If the leak is not serious enough to result in a total loss of prime, the pump may operate at a reduced head or capacity due to the air mixing with the water. This causes the water to be lighter than normal and reduces the efficiency of the energy transfer process.

- **High Efficiency Range Is Narrow:** As we have seen in the pump characteristic curves, a centrifugal pump's efficiency is directly related to the head capacity of the pump. The highest performance efficiency is available for only a very small section of the head-capacity curve. When the pump is operated outside of this optimum range, the efficiency may be greatly reduced.
- **The Pump May Run Backwards:** The centrifugal pump does not have the built-in capability to prevent flow from moving through the pump in the opposite direction (i.e., from discharge side to suction). If the discharge valve is not closed or the system does not contain the proper check valves, the flow that was pumped from the supply tank to the discharge point will immediately flow back to the supply tank when the pump shuts off. This results in increased power consumption, because of the frequent startup of the pump to transfer the same liquid from supply to discharge. (Note: It may be very difficult to determine if this is occurring because the pump looks and sounds like it is operating normally when operating in reverse.)
- **Pump Speed Is Not Easily Adjusted:** Centrifugal pump speed usually can't be adjusted without the use of additional equipment (such as speed-reducing or speed-increasing gears or special drive units). Because the speed of the pump is directly related to the discharge capacity of the pump, the primary method available to adjust the output of the pump other than a valve on the discharge line is to adjust the speed of the impeller. Unlike some other types of pumps, the delivery of the centrifugal pump can't be adjusted by changing some operating parameter of the pump.

3.7 WATER/WASTEWATER APPLICATIONS

As stated several times earlier, the centrifugal pump is probably the most widely used pump available at this time, because of its simplicity of design and the fact that the pump can be adjusted to suit a multitude of applications. Proper selection of the pump components (impeller, casing,

TABLE 3.3. Centrifugal Pump Applications in Water Systems.*

Application	Function
Low service	To lift water from the storage to treatment processes or from storage to filter-backwashing system
High service	To discharge water under pressure to distribution system
Booster	To increase pressure in the distribution system or to supply elevated storage tanks
Well	To lift water from shallow to deep wells and discharge it to the treatment plant, storage facility, or distribution system
Sampling	To pump water from sampling points to the laboratory or automatic analyzers
Sludge	To pump sludge from sedimentation facilities to further treatment or disposal

*Adapted from AWWA, 1996, p. 238.

etc.) and construction materials can produce a centrifugal pump capable of transporting materials ranging from coal or crushed stone slurries to air (centrifugal blowers used for aeration). To attempt to list all of the various applications for the centrifugal pump in water and wastewater treatment would exceed the limitations of this handbook. Therefore, the discussion of pump applications is limited to those that occur most frequently in water/wastewater treatment applications.

Water applications of the centrifugal pump are listed in Table 3.3. Wastewater applications of the centrifugal pump are listed in Table 3.4.

TABLE 3.4. Centrifugal Pump Applications in Wastewater Systems.

Application	Function
High-volume pumping	Generally, low speed, moderate head, vertically shafted, centrifugal pumps are used for high-volume capacity.
Non-clog pumping	Specifically designed centrifugal pumps using closed impellers with, at most, two to three vanes. Usually designed to pass solids up to 3 inches in diameter.

TABLE 3.4. (continued).

Application	Function
Dry pit pump	Depending on the exact application, may be either a large volume or non-clog pump. Located in a dry pit that shares a common wall with the wet well, this pump is normally placed in such a position as to ensure that the water level in the wet well is sufficient to maintain the pump's prime.
Wet pit or submersible pump	Usually a non-clog-type pump that can be submerged, together with its motor, directly in the wet well. In a few instances, the pump may be submerged in the wet well while the motor remains above the water level. In these cases, the pump is connected to the motor by an extended shaft.
Underground pump stations	Using the wet well-dry well design, the pumps are located in an underground facility. Wastes are collected in a separate wet well, then pumped upward and discharged into another collector line or manhole. This system normally uses a non-clog-type pump and is designed to add sufficient head to the waste flow to allow gravity flow to the plant or the next pump station.
Recycle or recirculation pumps	Because the liquids being transferred by the recycle or recirculation pump normally do not contain any large solids, the use of the non-clog-type centrifugal pump is not always required. A standard centrifugal pump may be used to recycle trickling filter effluent, return activated sludge, or digester supernatant.
Service water pumps	The plant effluent may be used for many purposes. It can be used to clean tanks, water lawns, provide the water to operate the chlorination system, and backwash filters. Because the plant effluent used for these purposes is normally clean, the centrifugal pumps used closely parallel those units used for potable water. In many cases, the double suction, closed impeller or turbine pump will be used.

REFERENCES

AWWA, *Water Transmission and Distribution,* 2nd ed. Denver: American Water Works Association, 1996.

Basic Maintenance Training Course, North Syracuse, New York: Onondaga County Department of Drainage and Sanitation, 1984.

Card, O. S., *Seventh Son.* New York: Tor Books, 1987.

Cheremisinoff, N. P. and Cheremisinoff, P. N., *Pumps/Compressors/ Fans: Handbook.* Lancaster, PA: Technomic Publishing Co., Inc., 1989.

Garay, P. N., *Pump Applications Desk Book.* Lilburn, GA: The Fairmont Press, Inc., 1990.

Lindeburg, M. R., *Civil Engineering Reference Manual,* 4th ed. San Carlos, CA: Professional Publications, Inc., 1986.

Metcalf & Eddy, *Wastewater Engineering: Collection and Pumping of Wastewater.* Tchobanoglous, G. (ed.). New York: McGraw-Hill Book Company, 1981.

Self-Test

3.1 Name three major components of the centrifugal pump.

3.2 Briefly explain how the centrifugal pump operates.

3.3 List three advantages of the centrifugal pump.

3.4 List three disadvantages of the centrifugal pump.

3.5 Explain how the volute casing causes a change from velocity head to pressure head.

3.6 Identify the components indicated in the drawing below. Compare the numbers on the drawing to the list provided.

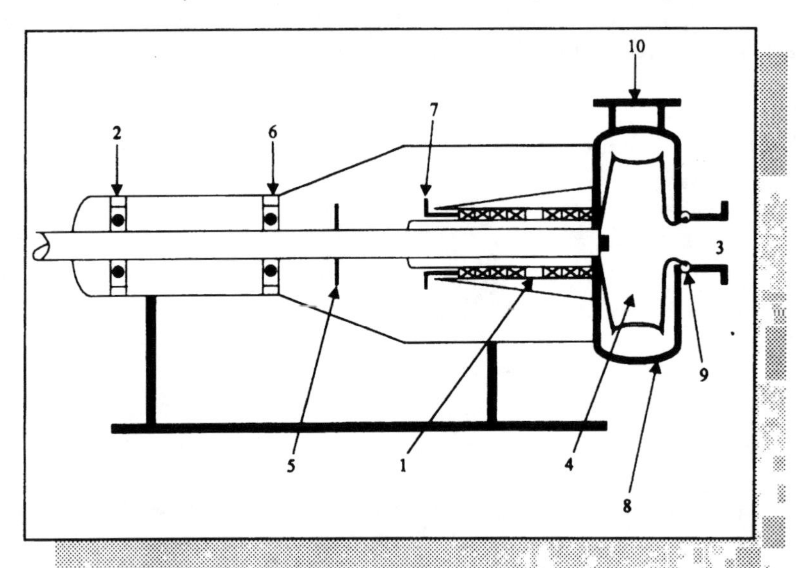

For Question 3.6

________ a. Volute case	________ f. Thrust bearing
________ b. Packing gland	________ g. Suction
________ c. Impeller	________ h. Discharge
________ d. Stuffing box	________ i. Radial bearing
________ e. Slinger ring	________ j. Impeller wear ring

Centrifugal Pump Components

All centrifugal pumps utilize but one pumping principle: the impeller rotates the water at high velocity, building up a velocity head.[15]

TOPICS

Casing
Impeller
Wear Rings
Shaft, Sleeves, and Couplings
Stuffing Box and Seals
Bearings

4.1 INTRODUCTION

In the chapter's opening statement, Garay succinctly points out the very basic operating principle of a centrifugal pump. On the surface, the simplicity of this statement points to the simplicity of the centrifugal pump itself. However, we must keep in mind that, although it is a "simple" hydraulic machine, it is also a composite of several major components, that should be familiar to those water/wastewater maintenance operators who must perform routine maintenance on the pump. Earlier, we briefly touched on the components making up the simple centrifugal pump. In this chapter, we describe each of the centrifugal pump's major components (i.e., the casing, impeller, shafts and couplings, stuffing boxes, and bearings) in greater detail, including their construction and their function.

4.2 CASING

The basic component of any pump is the housing or *casing*, which directs the flow of water into and out of the pump. The housing sur-

[15]From Garay, P. N., *Pump Application Desk Book.* Lilburn, GA: The Fairmont Press, Inc., p. 22, 1990.

Key Terms Used in This Chapter

AXIAL LOAD	A load parallel to the shaft
CASING (VOLUTE)	The enclosure surrounding the pump impeller, shaft, and stuffing box
DISCHARGE OUTLET	The passage through which the pump discharges water to the piping system
IMPELLER	The part of the pump that supplies energy to the water to give it velocity and momentum
IMPELLER HUB	The portion of the impeller that mounts on the shaft
IMPELLER VANES	Devices that direct the flow of water within the pump
LANTERN RING	A device used to distribute sealing fluid within a stuffing box
MECHANICAL SEAL	A molded seal held in place by springs or other constant-pressure devices
RADIAL LOAD	A load perpendicular to the shaft
STUFFING BOX	A sealing area that is manually packed and adjusted
SUCTION INTAKE	The passage through which water enters the pump

rounding the impeller of a centrifugal pump is called the *volute case.* The word volute is used to describe the spiral-shaped cross-section of the case as it wraps around the impeller; that is, the pump casing gets larger as it nears the discharge point [see Figure 4.1(a)].

IMPORTANT ***Important Point:*** *If the pump casing was the same size all the way around, the water flow would be restricted, and the pump could not develop its rated capacity.*

In addition to enclosing the impeller, the volute case is cast and machined to provide the seat for the impeller wear rings. The volute case also includes suction and discharge piping connections. In the volute

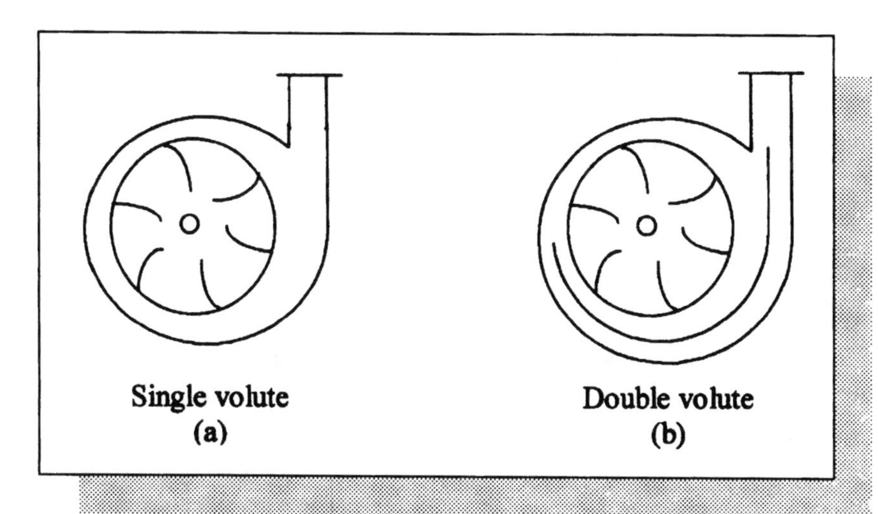

Figure 4.1
Types of volutes.

pump shown in Figure 4.1(a) the pressure against the impeller is unbalanced, resulting in an unbalanced load, which is taken by the bearings supporting the impeller shaft. The pump is designed for a radial load on the bearings, and as long as the pump performs at conditions not too far from the design point, the radial loading is accommodated. However, if the pump is operated at less than 30 percent or more than 120 percent of design capacity, the radial load increases drastically, causing early failure of the bearings. More significantly, the unbalanced load can cause excessive shaft deflection in areas of fine running clearances and eccentric loading of mechanical seals, resulting in leakage.[16]

To reduce this unbalanced load problem, double volute diffuser casings such as the one shown in Figure 4.1(b) are used. In double volute casings, while the pressures are not uniform at partial capacity operation, the resultant forces for each 180°-volute section oppose and balance each other. That is, the double volute incorporates a flow splitter into the casing that directs the water into two separate paths through the casing. The contour flow of the splitter follows the contour of the casing wall 180° opposite. Both are approximately equidistant from the center of the

[16]Renner, D., *Hands-on Water/Wastewater Equipment Maintenance.* Lancaster, PA: Technomic Publishing Co., Inc., p. 181, 1999; Garay, P. N., *Pump Application Desk Book.* Lilburn, GA: The Fairmont Press, Inc., p. 51, 1990.

impeller; thus, the radial thrust loads acting on the impeller are balanced and greatly reduced.

The volute casing can be classified as either solid or split casing.

4.2.1 SOLID CASING

The *solid volute casing* is designed as a single piece of casting with a top or bottom opening to install or work on the impeller and wear rings. An example of this type of pump is the end suction centrifugal pump shown in Figure 4.2. In this design, the bottom section of the volute is bolted to the intake or suction line, and the top opening of the volute is covered with the shaft assembly. End suction pumps are easy to recognize because the suction and discharge nozzles are usually at 90° angles to each other. To help simplify internal inspections of these pumps, without disassembly, the volute case is often equipped with removable inspection plates.

IMPORTANT *Note: A modification of the solid volute centrifugal pump is the back pullout pump. In this type, the volute is connected to the suction and discharge piping. The pump itself is pulled out from the*

Figure 4.2
Solid cased, end suction centrifugal pump.

back of the volute. This modification enables the operator to inspect or work on the pump without having to disconnect any piping or dismantle the pump.

4.2.2 SPLIT CASINGS

The *split case* pump uses two or more sections fastened together to form the volute case (see Figure 4.3). Depending on the direction of the split, these pumps can be classified as axially or radially split. Axially split casings are split parallel to the pump shaft. When half the casing is removed, the length of the shaft and the edge of the impeller are visible. Radially split casings split perpendicular to the pump shaft. When these pumps are opened, a cross section of the shaft and the face or back of the impeller are visible. The suction and discharge for split case pumps are in the same half, parallel to each other but on opposite sides (this arrangement allows half of the casing to be removed for easy inspection of the interior without disturbing the piping, bearings, and/or shaft assembly).

The casings on volute-type centrifugal pumps can be modified further to increase the volume of water handled and/or the pressure obtained.

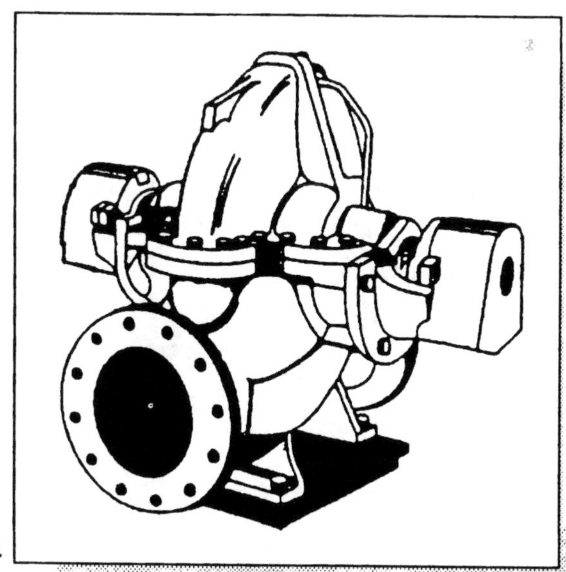

Figure 4.3
Split case pump.

Most of the pumps we discuss in this handbook have a single impeller and a single suction. Volute pumps may also be multi-stage pumps (i.e., having two or more impellers and a corresponding number of volute cases). The discharge of the first volute serves as the suction of the second and so on. Multi-stage pumps with two stages are capable of obtaining twice the pressure of a comparable single-stage pump. However, the volume or quantity of flow remains unchanged. To change the quantity of flow, the volute suction size is increased. Instead of a single suction, some pumps are designed with double suctions. A double-suction, single-stage pump can discharge twice the volume of water discharged by a single-suction, single-stage pump with both discharging at equal pressures. To increase both the volume and pressure, a double-suction, multi-stage pump could be used.

4.3 IMPELLER

The heart (and thus the most critical part) of a centrifugal pump is the *impeller.* Moreover, the impeller's size, shape, and speed determine the pump's capacity. Although there are several designs for impellers, each transfers the mechanical energy of the motor to velocity head by centrifugal force. The central area of the impeller is called the **hub.** The hub is machined so the impeller can be attached securely to the pump shaft. Surrounding the hub is a series of rigid arms, called vanes, that extend outward in a curved shape (see Figure 4.4). The vanes throw the water into the volute case, causing an increase in the velocity of the water. Depending on the type of impeller, the impeller vanes will vary in thickness, height, length, angle, and curvature. To increase the impeller efficiency and strengthen its construction, some impellers are enclosed by sidewalls called shrouds.

Impellers can be classified as:

- semi-open
- open
- closed

4.3.1 SEMI-OPEN IMPELLER

Semi-open impellers [see Figure 4.4(a)] have only one shroud on the back of the impeller that covers the hub and extends to the edge of the vanes. That is, when seen from the back, the shroud forms a complete circle. This feature allows the vanes to be thicker and less likely to be damaged by collision with solids or debris. The face of this impeller is left open. The shroud, besides adding structural stability, increases the efficiency of the impeller.

Semi-open impellers are most often used for pumping liquids with medium-sized solids but they are capable of handling high solids concentrations. They are capable of pumping high volumes of liquid at low pressures. The size of the solids that an open or semi-open impeller can pump depends on the closeness of the impeller to the suction side of the volute case. The distance can vary from 0.015 inches to several inches.

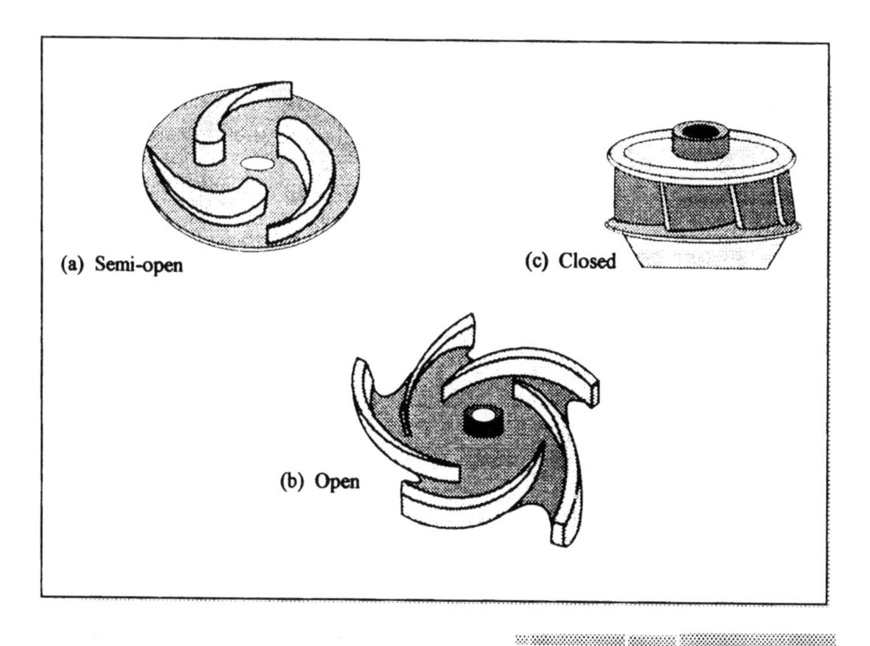

Figure 4.4
Centrifugal pump impellers.

4.3.2 OPEN IMPELLER

Open impellers are designed with vanes (curved blades) that extend from the hub with no top or bottom shroud [see Figure 4.4(b)]. Some open impellers do, however, have a partial bottom shroud to strengthen the impeller vanes. Open impellers are used to pump water with large-sized solids and/or water with high solids concentrations. They are generally capable of pumping high volumes of water at low pressures. Open impellers are more easily damaged than the semi-open or closed impeller due to the exposed vanes.

4.3.3 CLOSED IMPELLER

The *closed impeller* has a shroud on both the front and back [see Figure 4.4(c)]. This arrangement leaves only the suction eye and the outer edge of the impeller open. With both shrouds, the impeller is quite strong and is able to maintain good pumping efficiency. The closed impeller is generally used for pumping clean water or clear wastewater. The size of the solids handled by a closed impeller pump will vary as the width of the vanes increases or decreases from one impeller to another. In contrast to open and semi-open impellers, closed impeller pumps can handle varying volumes of water and can develop very high pressures.

For use in wastewater pumping and to maintain the high level of pumping efficiency and at the same time pump varying volumes of raw wastewater at high pressures, non-clogging closed impellers were developed. These non-clogging impellers have large internal openings, and the distance between the shrouds is expanded so large solids will pass through them (see Figures 4.5 and 4.6). Normally, a wastewater pump will be designed to allow passage of solids up to 3" in size.

Important Point: *Impellers may also be classified as to whether they are single or double suction. Single-suction impellers have their flow coming into the impeller from one side only. Double-suction impellers have flow entering from both sides; therefore, they have two suction eyes instead of one. A double-suction impeller does not*

Figure 4.5
A "well-worn" non-clogging closed impeller used in wastewater treatment.

Figure 4.6
A type of non-clogging closed impeller used in pumping waste-water solids.

increase the pressure obtained by the pump yet it does double the amount of water being pumped. As was discussed in Section 4.2 (casing), some pumps use combinations of impellers and suction piping to increase pumping capabilities without adding additional pumps and drive units.

4.4 WEAR RINGS

As the impeller of a centrifugal pump spins, it creates a low-pressure zone on the suction side of the impeller, by the impeller eye. As the water is thrown off the impeller vanes by centrifugal force, it creates a high-pressure zone inside the volute case. If the impeller and volute case were not matched so that the clearance between them was small, water from the high-pressure zone in the volute would flow to the low-pressure zone in the eye of the impeller and be repumped. To prevent this from occurring (i.e., to provide physical separation between the high- and low-pressure sides), a flow restriction must exist between the impeller discharge and suction areas. Wear rings accomplish this restriction of flow (referred to as recirculation). The wear rings prevent permanent damage to the volute case and impeller. The most widely used materials for wear rings are bronze or brass alloys and are replaceable items. Bronze exhibits good resistance to corrosion and abrasion, with excellent casting and machining properties. Wear rings may be installed in the front and the back of the volute and on the impeller itself (see Figure 4.7).

When a wear ring is mounted in the case of a pump, it is called a casing ring. When it is mounted in the suction area of the pump, it is called

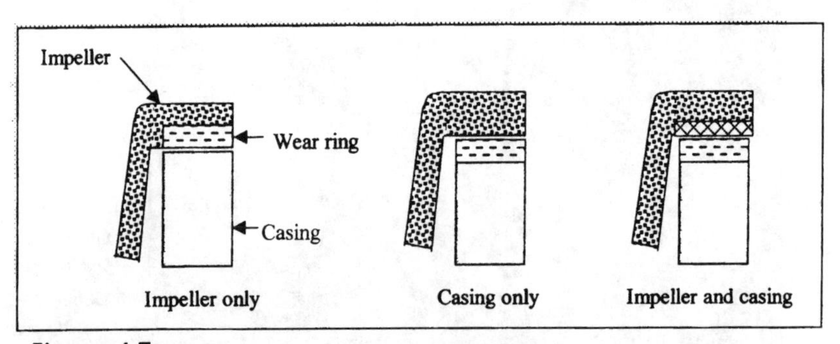

Figure 4.7
Wear ring arrangements.

a suction ring. If the suction-head ring is the only wear ring installed, both the ring and impeller must be replaced at the same time to maintain the proper clearance. If the pump has wear rings mounted on the impeller (impeller rings) and casing (suction head rings), only the wear rings will need to be replaced. The impeller can be reused provided there is no other damage. Pumps with casing or suction head rings and impeller rings have double ring construction. They have both a stationary and a rotating ring.

Wear rings can also be installed at the stuffing box, in which case they are called stuffing-box cover rings. Regardless of where the rings are installed, they are usually secured with set or machine screws along with some kind of locking device. This stops them from turning and wearing against their volute case seat. However, an exception to this is the impeller wear rings that sometimes are installed as pressure fit or shrink fit pieces instead of using screws to secure them.

The clearance between the wear rings should be checked whenever a pump is opened for routine inspection or maintenance. Check the manufacturer's technical manual for proper clearance data.

IMPORTANT

Note: *If a pump does not have wearing rings, worn parts must be replaced or rebuilt. On some small pumps, parts replacement may be inexpensive. On large pumps, however, the cost of wearing rings is far less than the cost of replacing the worn parts.*

4.5 SHAFTS, SLEEVES, AND COUPLINGS

Important to the operation of any centrifugal pump and drive unit is the shafting, sleeves, and couplings used to connect the drive unit to the pump.

4.5.1 SHAFTING

Shafting for a centrifugal pump consists of a main pump shaft plus possible intermediate shafts for connecting drive units where the pump and drive are separated from each other. The main pump shaft (see Figure 4.8) is a solid shaft constructed of high-quality carbon or stainless steel to increase its resistance to wear and corrosion. (Note: Although corrosion-resistant materials are expensive, it is usually good practice to install a

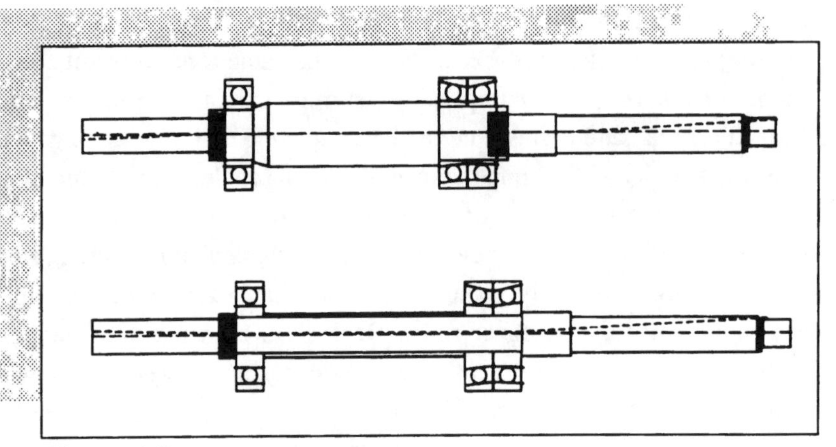

Figure 4.8
Pump shafts for centrifugal pumps.

Figure 4.9
Bearing press.

high-quality shaft despite the higher initial cost.) The shaft supports the rotating parts of the pump and transmits mechanical energy from the drive unit to the pump impeller. A common method used to secure the impeller to the shaft on double-suction pumps involves using a key and a very tight fit. Because of the tight fit, an arbor press (see Figure 4.9) or gear puller is required to remove an impeller from the shaft.[17] In end-suction pumps, the impeller is mounted on the end of the shaft and held in place by a key nut.

The shaft is designed to withstand the various forces acting on it and still maintain the very close clearances needed between the rotating and stationary parts. Although the shaft is of solid construction, care must be exercised when working on or around it. Slight dents, chips, or strains are capable of causing misalignment or bending of the shaft.

Closed-coupled pumps (see Figure 4.10) that have the casing mounted directly onto the drive motor have different shaft designs and construction features than frame-mounted pumps. Simply, in this pump, the impeller and the drive unit share a common shaft. That is, the shaft that supports the impeller is actually the motor shaft that has been extended into the pump casing (see Figure 4.10).

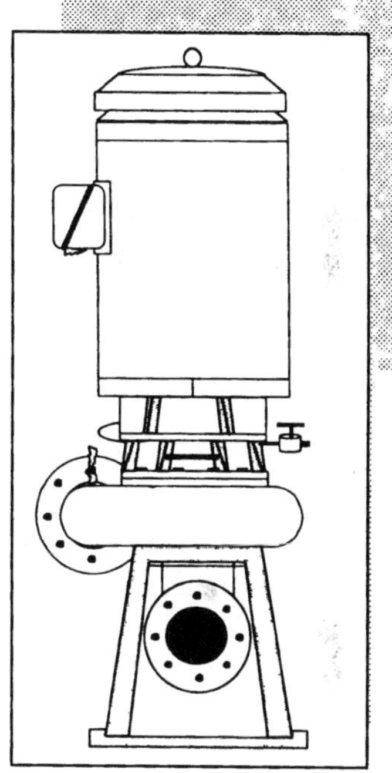

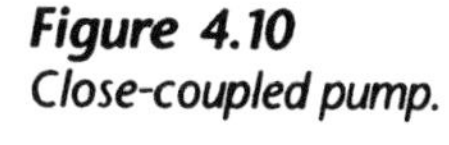

Figure 4.10
Close-coupled pump.

4.5.1.1 INTERMEDIATE SHAFTS

Not all pump drive systems are designed so that the unit and the pump can be coupled directly (see Section 4.5.3). In many cases, distances from several inches to 100 feet separate the drive and pump units. In these situations, *intermediate shafts* are used to transfer energy. What is required may vary from spacers to floating or rigid shafts to flexible drive shafts.

[17]AWWA, *Water Transmission and Distribution*, 2nd ed. Denver: American Water Works Association, p. 381, 1996.

One way to bridge the shaft separation is to use a one-piece flanged tubular spacer. A flanged tubular spacer is used for gaps up to several feet. Beyond that, the cost of manufacturing the spacer is prohibitive. The spacer is connected to the flanges of the coupling and bridges the gap between the shafts.

Floating shafts accomplish the same task as a spacer; however, they are constructed differently. Floating shafts are made by attaching a flange to a piece of solid or tubular shafting by a mechanical key or by welding. This construction is less expensive than the one-piece spacer, yet, like the spacer, the flanges on the ends of the shaft connect directly to the coupling flanges. Long sections of floating shaft need to be supported by line bearings at intermediate supports or floors. Floating shaft arrangements are widely used on horizontal pump applications and are especially common on vertical systems. Axial thrust loads in a floating shaft system are compensated for by the pump's thrust bearing; therefore, the couplings between the shaft segments can be of the flexible type.

Some pumps are designed with only a single line bearing associated with the pump. In these situations, the axial thrust load is taken up by the thrust bearing in the drive unit. In a pump system of this kind, rigid intermediate shafts and couplings are required. The pump and drive unit couplings and those on intermediate shafts must be rigid if the axial thrust is to be transmitted to the drive unit. Bearings associated with this system must only provide lateral support; in other words, they are line bearings.

In many vertical and horizontal pump applications, flexible drive shafts are used as intermediate shafts. In these applications, universal joints with tubular shafting can be substituted for flexible couplings when

- there is a need for critical alignment
- the space to be spanned is considerable
- there is a possible need to permit large amounts of motion between pump and drive unit

Flanges are used to fit the pump and drive unit shafts to the universal joints. These joints are splined to allow movement of the intermediate shafts; therefore, pump thrust has to be taken up by combination pump thrust and line bearing. Intermediate bearings are required to steady the shafts. However, these bearings do not take on any radial loads because these are taken by the universal joints.

4.5.2 SLEEVES

Most centrifugal pump shafts are fitted with brass or other nonferrous metal sleeves. *Sleeves* protect the shaft from erosion and corrosion and provide a wearing surface for packing or a place to mount the mechanical seals. (Note: Permitting the sleeves to take the wear from the packing rather than the shaft keeps maintenance costs and time to a minimum, compared with the replacement of a shaft.[18]) Shaft sleeves serving other functions are given specific names to indicate their purpose. For example, a shaft sleeve used between two multi-stage pump impellers in conjunction with the interstage bushing to form an interstage leakage point is called an *interstage* or *distance sleeve.*[19]

4.5.3 COUPLINGS

To allow energy transfer from the drive unit or motor to the pump, these units must be connected; that is, they must be coupled. Renner[20] points out that the primary duty of a *coupling* is to transmit motion and power from one source (drive unit) to another (the pump). To accomplish this, couplings have to meet three basic design requirements (see Figure 4.11):

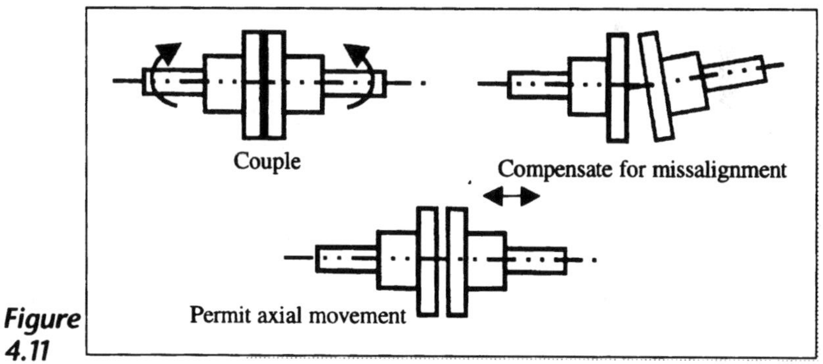

Figure 4.11
Coupling requirements.

[18]Renner, D., *Hands-on Water/Wastewater Equipment Maintenance.* Lancaster, PA: Technomic Publishing Co., Inc., p. 184, 1999.

[19]Karassik, I. J., Centrifugal Pump Construction, *Pump Handbook,* Karassik, I. J. et al. (eds.). New York: McGraw-Hill, pp. 2–71, 1976.

[20]Renner, D., *Hands-on Water/Wastewater Equipment Maintenance.* Lancaster, PA: Technomic Publishing Co., Inc., p. 121, 1999.

- Couple two rotating shafts together to transmit power and motion from one machine to another.
- Compensate for any misalignment between the two rotating members.
- Allow for axial or end movement between the coupled shafts.

4.5.3.1 COUPLING CATEGORIES

Pump couplings are grouped into two broad categories:

- rigid couplings
- flexible couplings

4.5.3.1.1 RIGID COUPLINGS

Rigid couplings, like all couplings, are used to transfer energy. Their rigid construction, however, allows for no shaft misalignment. If misalignment does exist when a rigid coupling is used, the coupling and pump and motor bearings will wear very quickly. Two commonly used types of rigid couplings are the flange and split couplings.

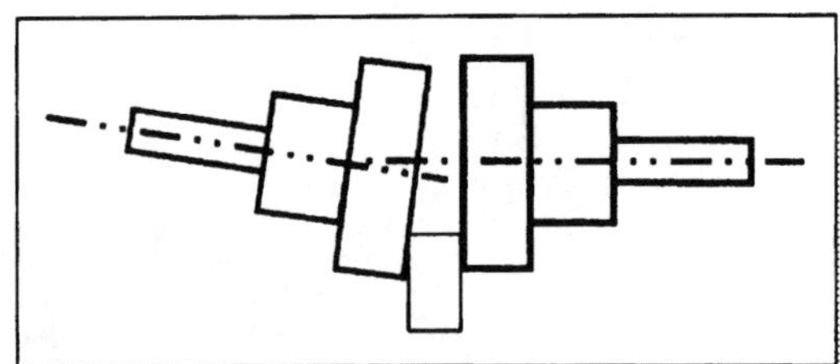

Figure 4.12
Angular misalignment.

IMPORTANT

Important Point: Coupling Misalignment*—When connecting two shafts, it is possible to have three different kinds of misalignment.* ***Angular*** *misalignment is where the flat surfaces of the ends of the shafts are not parallel with each other (see Figure 4.12).* ***Parallel*** *misalignment (see Figure 4.13) is where the centers of the two shafts are not directly over each other, and the third type of misalignment is a combination of both angular and parallel.*

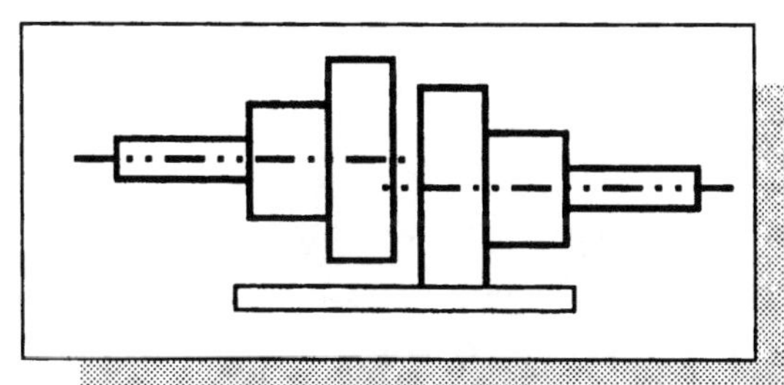

Figure 4.13
Parallel misalignment.

4.5.3.1.1.1 FLANGED COUPLING

A properly flanged coupling consists of two flanges; one attached to each shaft as shown in Figure 4.14. Each flange has a replaceable center bushing with a keyed slot. The keyed slot matches the shaft key, and the bearing can be changed to match the different shaft diameters. When properly installed, the flanges are held together by bolts. The bolts, however, do not function in energy transfer. The frictional force of the two flange faces touching transfers energy from one shaft to another.

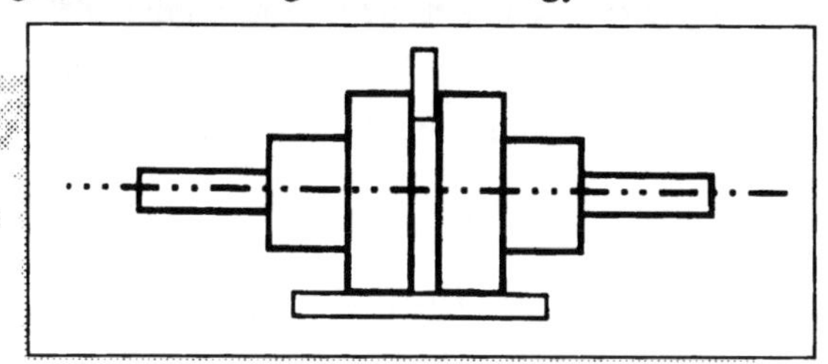

Figure 4.14
Two flanges correctly aligned.

4.5.3.1.1.2 SPLIT COUPLING

The *split coupling* is a tubular coupling that is split axially and held together and around the shaft by bolts. One half of the coupling is keyed and matches up with the keys of the two shafts. The split coupling offers the advantage of easy installation and removal. It also allows a certain amount of impeller adjustment due to its long tubular shape.

IMPORTANT

***Note:** Rigid couplings find their widest use on vertical-mounted pumps.*

4.5.3.2 FLEXIBLE COUPLINGS

Flexible couplings allow the transfer of energy and compensate for small amounts of shaft misalignment. Flexible couplings are mechanically flexible or materially flexible. There are several types available.

Note that the type of flexible coupling used for each pump application varies with the horsepower of the drive unit, speed of rotation, shaft separation, amount of misalignment, cost, and reliability requirements. The few discussed previously are not the only ones available. If a coupling needs to be replaced at your facility, the manufacturer's literature or representative should be consulted.

4.5.3.2.1 MECHANICALLY FLEXIBLE COUPLINGS

Mechanically flexible couplings compensate for misalignment between two connected shafts by providing internal clearances within the design of the coupling. Examples of this are chain coupling and gear coupling.

4.5.3.2.1.1 CHAIN COUPLING

A *chain coupling* consists of a gear attached to each shaft with a double-width chain wrapped around the two gears. The spacing between the faces of the gears and the flexibility in the chain compensate for misalignment. This type of coupling is limited to low speed equipment and should be surrounded by a housing for safety reasons. Lubricant is often placed inside the housing to reduce friction and extend the life of the coupling.

4.5.3.2.1.2 GEAR COUPLING

The *gear coupling* consists of a gear assembly keyed onto each shaft and surrounded by a housing with corresponding internal gears. The self-adjusting gear assemblies compensate for misalignment. Like the chain coupling, the housing of the gear coupling should have a clean supply of lubricant to reduce wear and extend coupling life.

4.5.3.2.2 MATERIALLY FLEXIBLE

Materially flexible couplings rely on flexible elements designed into the coupling to compensate for misalignment. Examples of this type of coupling include the jaw coupling, the flexible disc and the flexible diaphragm coupling.

4.5.3.2.2.1 JAW COUPLING

The *jaw coupling* is one of the most common and least expensive of the materially flexible couplings. It consists of two flanges, one keyed on each shaft, with each flange having three triangular teeth. An elastic piece of rubber, the spider, separates the flanges and teeth and helps transfer the energy. The jaw coupling compensates for all types of misalignment, but can contribute to vibration in well-aligned units.

4.5.3.2.2.2 FLEXIBLE DISC COUPLING

The *flexible disc coupling* consists of two flanges similar to those on the flange coupling that are keyed, one on each shaft. Each flange has pins protruding from it that pass through a flexible circular disc and into a slot in the other flange. The flexible disc compensates for any misalignment between the shafts. The flexible disc coupling can compensate up to 2° angular and 1/32" parallel misalignment.

4.5.3.2.2.3 FLEXIBLE DIAPHRAGM COUPLING

The *flexible diaphragm coupling* consists of two flanges, one keyed on each shaft with a rubber or synthetic diaphragm enclosing the space around the flanges. These couplings can handle up to 4° angular and 1/8" parallel misalignment.

4.6 STUFFING BOX AND SEALS

Sealing devices are used to prevent water leakage along the pump driving shaft. Shaft sealing devices must control water leakage without causing wear to the pump shaft. There are two systems available to accomplish this seal: the conventional stuffing box/packing assembly and the mechanical seal assembly.

4.6.1 STUFFING BOX PACKING ASSEMBLY

The *stuffing box* of a centrifugal pump is a cylindrical housing, the bottom of which may be the pump casing, a separate throat bushing attached to the stuffing box, or a bottoming ring. There are a number of different designs of stuffing boxes for pumps used in water/wastewater plants.

4.6.1.1 PACKING GLAND

At the top of the stuffing box is a *packing gland* (see Figure 4.15). The gland encircles the pump shaft or shaft sleeve and is cast with a flange that slips securely into the stuffing box. Stuffing box glands are manufactured as a single piece split in half and held together with bolts. The advantage of the split gland is the ability to remove it from the pump shaft without dismantling the pump.

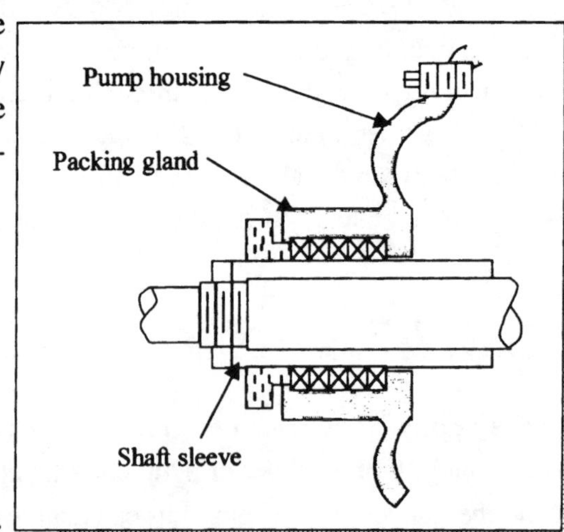

Figure 4.15
Solid packed stuffing box.

4.6.1.2 PACKING MATERIAL

The sealing material placed inside the stuffing box is a *packing material.* In conventional pumping systems, the stuffing box/packing system is generally used to seal the pump. The type of packing used varies from operation to operation depending on the type of service for which the pump is designed.

The materials most commonly used for packing in pumps employed in water/wastewater operations include flax or cotton. However, there are a number of different kinds of packing, including metallic foil or synthetic substances like Teflon®, that are used or recommended for use to meet varying temperature, pressure, and/or liquid composition conditions. Generally, the raw materials are woven or braided to make continuous square-shaped strands; but other patterns, like circular braided strands, are also available. The strands are sometimes wire reinforced and usually contain graphite or an inert oil lubricant that helps bond the braided strands together and reduce the friction between the stationary packing and the rotating shaft.

Packing is purchased in either continuous rope-like coils, with a square cross section, or as pre-formed die-molded rings. When the rope-like packing is used, it is cut in sections to make up the number and size of the rings required. Some maintenance personnel prefer, where possible, to use the die-molded rings because they ensure an exact fit to the shaft or shaft sleeve and the inside wall of the stuffing box-die-molded rings ensure a uniform packing density throughout the stuffing box. Precut rings are generally available in exact sizes and numbers for repacking most pumps.

4.6.1.3 LANTERN RINGS

As stated earlier, the purpose of the stuffing box/packing assembly is to seal the opening where the pump shaft passes into the pump. This prevents air from leaking into the pump or the pumped water from leaking out (except for a controlled amount). When either of these conditions exist, pump efficiency decreases. To seal the opening, packing is placed inside the stuffing box, and the packing gland applies pressure to it. This squeezes the packing and forces it to fill the area between the shaft and the stuffing box wall. This seals the area. However, when operating, the friction and heat build-up between the stationary packing and the rotating shaft destroys the packing. The packing, although it is lubricated, quickly becomes worn and hard. This destroys the seal and possibly damages the shaft. To prevent this from occurring, a lantern ring is placed in the stuffing box along with the packing, directly across from an opening in the stuffing box.

The *lantern ring* or seal cage (see Figure 4.16) is a circular brass (metallic) or plastic ring, split into equal halves and placed around the pump shaft or shaft sleeve inside the stuffing box. The lantern ring has an I-beam construction and holes drilled through it.

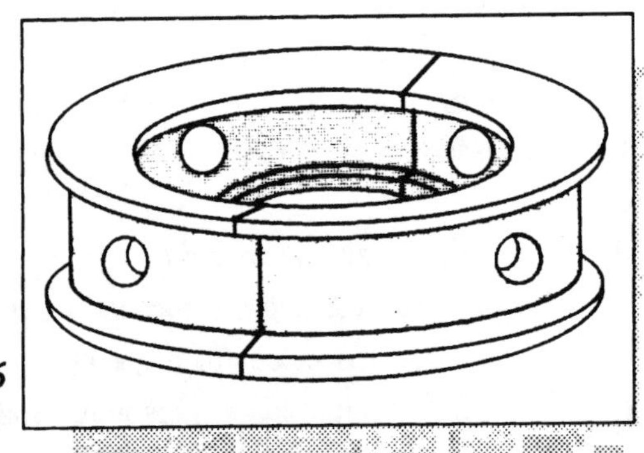

Figure 4.16
Lantern ring.

The lantern ring allows sealing liquid to flow around and through the lantern ring to lubricate and cool the packing and aid in sealing the pump. The location of the lantern ring inside the stuffing box is determined when the pump is manufactured. It is very important when repacking a pump that the lantern ring be replaced in the proper sequence along with the packing to ensure the proper distribution of the sealing liquid.

Experience has demonstrated that it is possible to order the same type of pump from a manufacturer with the lantern ring located in different positions. Having the lantern ring located closer to the inner portion of the pump diverts greater quantities of sealing liquid into the pump and therefore, helps to keep the pumped material out of the stuffing box more effectively. This could be very important when pumping gritty wastewater or sludges. These liquids can be very abrasive on the pump shaft, sleeve, or packing.

4.6.1.3.1 SEALING LIQUID

The *sealing liquid* piped into the stuffing can be fed from internal or external sources. When the water being pumped is clean and clear and will not damage the packing, the sealing liquid can be fed from the discharge side of the pump. This is done by either external or internal piping. When the material being pumped is abrasive or would damage the packing, an external source is used as a sealing liquid. The external sealing liquid is then pumped into the stuffing box by a small seal pump. (Note: The sealing liquid for an externally sealed pump is generally clean water; however, some installations use oil seals.)

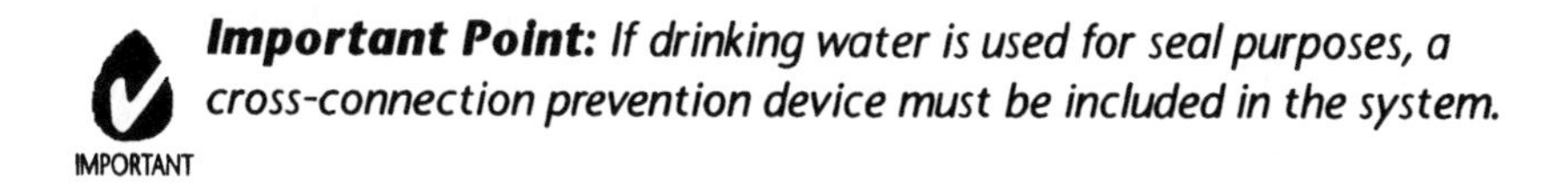

Important Point: *If drinking water is used for seal purposes, a cross-connection prevention device must be included in the system.*

In summary, the rings of packing help seal the pump while the lantern ring, placed between the packing, helps to feed the sealing liquid under pressure into the stuffing box to form an air-tight seal of the pump. The sealing liquid for an externally sealed pump is generally clean water; however, some installations use oil seals.

4.6.2 MECHANICAL SEALS

Mechanical seals are used in many pumps to prevent water (or other liquid) leakage. Mechanical seals might be chosen over packing on a given application for three reasons. First, mechanical seals provide a better fluid seal than packing. Second, mechanical seals usually require less maintenance than packing. Third, mechanical seals can withstand higher pressure than stuffing boxes.[21]

[21]TPC, *Understanding the Operation of Pumps.* Buffalo Grove, IL: TPC Training Systems, p. 141, 1986.

IMPORTANT

***Note:** Even though mechanical seals have many advantages over the stuffing box and its inherent problems of maintenance and water leakage, when mechanical seals fail, the entire pump must be disassembled and the units completely replaced.*

Recall that when packing is used to seal a pump, the sealing surface is between the packing and the shaft or shaft sleeve. This surface is parallel with the shaft of the pump. However, if mechanical seals are used, the sealing surface is perpendicular to the pump shaft (see Figure 4.17). A mechanical seal consists of two rings with highly polished surfaces. These surfaces run against one another. One surface, the rotating element, is connected to the pump shaft, and the other is connected to the stationary portion of the pump (see Figure 4.17). The stationary element of a mechanical seal is generally spring loaded to ensure continuous contact between the two polished surfaces—this fine finish is important if the seal surfaces are to match and seal properly. Very small quantities of sealing liquid are allowed to flow across the faces in order to complete the seal, as well as lubricate and cool the faces. This cuts down on the wear and increases the life of the mechanical seal.

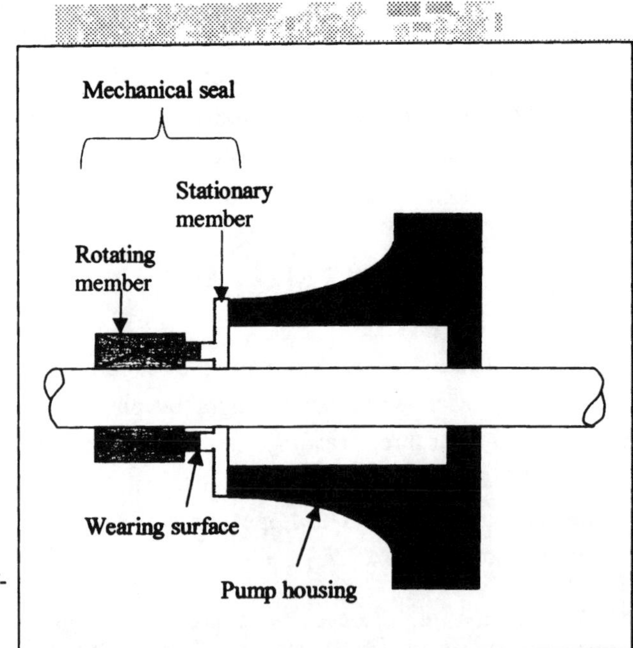

Figure 4.17
Typical mechanical seal.

New seals that are installed properly with the specified clearances have negligible leakage. The flow past the faces is almost non-existent. With time and wear, a small amount of leakage has to be expected. Once the leakage reaches excessive proportions, the entire seal has to be replaced.

4.6.2.1 MECHANICAL SEALS: SEALING POINTS

All mechanical seals have three primary sealing points. The first is the area between the stationary element and the seal housing. This area is sealed with regular gaskets or O-rings. The second is between the rotating element and the shaft. This is also sealed by O-rings. The third is between the polished faces. The seal water flow and the very close contact between these faces achieve this seal. To increase the life of the mechanical seal and to achieve a tight seal, the surface of the polished faces is made of dissimilar materials. For example, one face might be made of stainless steel while the other one is a synthetic Teflon® material.

4.7 BEARINGS

Simply, a *bearing* is a supporting surface that separates a stationary and a moving object. Though not always the case, generally, bearings are thought of as something that supports a rotating shaft. In a centrifugal pump, bearings maintain the alignment between the rotating parts (shaft and impeller) and the stationary parts (the case and frame). To maintain proper alignment, bearings must be able to work under varying loads. This load or force varies with the type of pump and the location of the bearing in the system. As the load or force varies, the bearings used to compensate also vary.

To maintain proper alignment, bearings work against radial and thrust loads. **Radial forces** push or pull the shaft and impeller out of alignment in directions perpendicular to the shaft (see Figure 4.18). **Thrust forces** push or pull the components out of alignment parallel to the shaft (see Figure 4.19). Bearings must maintain both the radial and axial positioning of the shaft and impeller.

Bearings used for centrifugal pumps can be described by their function, position, and type of construction. Bearings that maintain the radial positioning of the shaft or impeller are called *line bearings*. Those that maintain axial positioning are *thrust bearings*. In many cases, thrust bearings can play a dual role, providing both radial and axial positioning.

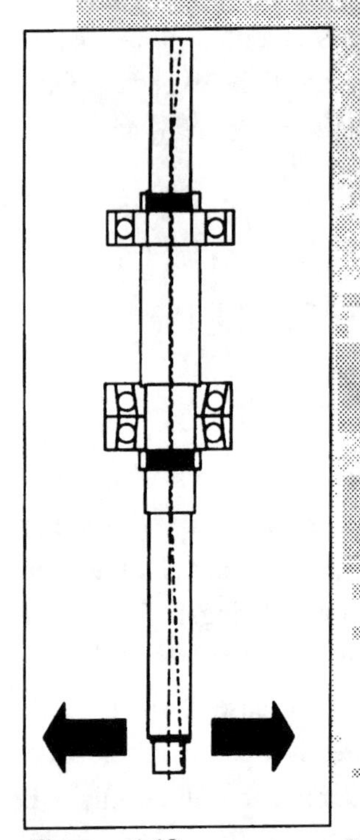

Figure 4.18
Radial forces.

In relation to the pump and motor or drive assembly, the terms inboard and outboard are used to describe the position of the bearings. In overhung impeller pumps, the inboard bearing is the one nearest the impeller. The outboard bearing is the one that is further away. Horizontal pumps have bearings at each end of the pump; the inboard bearing is located between the casing and the coupling while the outboard bearing is located on the other side of the casing. In horizontal pumps, thrust bearings are usually placed at the outboard end while line bearings are used on the inboard side. All centrifugal pumps have a thrust bearing, which is usually outboard, even though this bearing may also act as a radial bearing and may be a single-row antifriction bearing exactly like the outboard.[22]

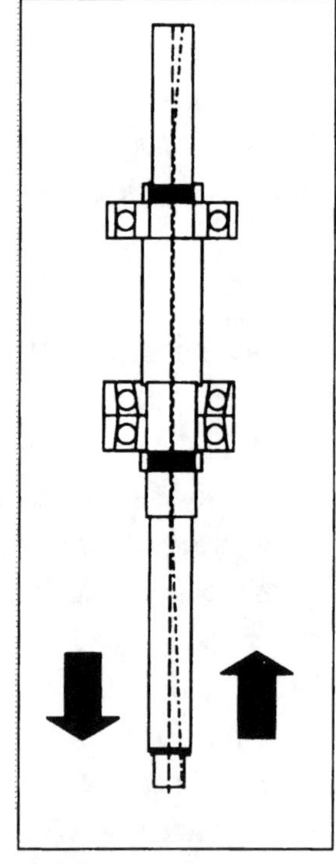

Figure 4.19
Thrust forces.

[22]Wahren, U., *Practical Introduction to Pumping Technology.* Houston: Gulf Publishing Company, p. 83, 1997.

4.7.1 APPLICATIONS AND TYPES OF PUMP BEARINGS

Nearly all types of bearings have been used on centrifugal pumps. The physical construction of the bearing helps to identify the type of bearing and its function. Bearings in general fall into two major categories—*sliding contact* bearings and *rolling contact* bearings. In the past, sliding contact bearings, also called *plain* or *sleeve bearings,* were once the bearing of choice; they now have been almost entirely replaced by rolling contact bearings for use on pumps. Anti-friction rolling contact-type bearings are most often used in centrifugal pumps. This type includes *ball bearings* and *roller bearings.*

Just as bearing applications vary, so do the types used. In centrifugal water/wastewater pumps, five types of bearings find widespread use. These are (1) self-aligning double-row ball bearings, (2) single- or double-row anti-friction ball bearings, (3) angular-contact ball bearings, (4) self-aligning, spherical roller bearings, and (5) single-row tapered roller bearings.

4.7.1.1 SELF-ALIGNING DOUBLE-ROW BALL BEARING

The self-aligning double-row bearing operates well against radial loads; however, it is only capable of withstanding very low thrust loads. For this reason, the self-aligning double-row ball bearing is ideally suited for line bearing applications. It operates well under heavy loads, high speeds, and long bearing spans as long as there is no end thrust loads.

4.7.1.2 SINGLE- OR DOUBLE-ROW ANTI-FRICTION BALL BEARING

The single-row (deep-groove) bearing is the most commonly used bearing on all but the larger centrifugal pumps. It can withstand radial

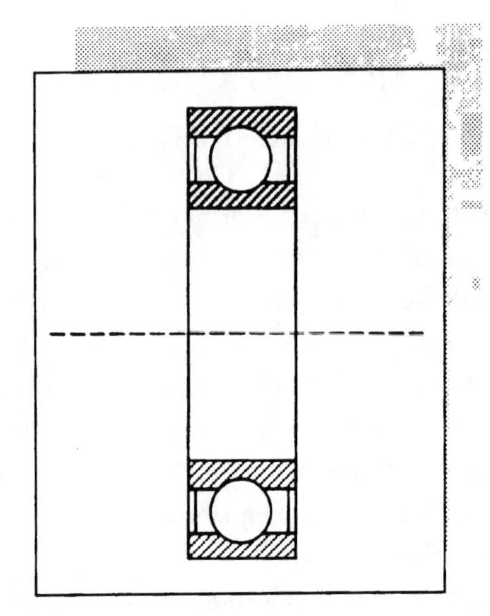

Figure 4.20
Single-row anti-friction ball bearing.

loads and reasonable amounts of thrust loading, and it operates well at high speeds. It does require careful alignment between the shaft and the bearing housing. A single-row anti-friction ball bearing is shown in Figure 4.20.

The double-row version of this same bearing (see Figure 4.21) has greater capacity to handle both radial and thrust loads. It is used when a single row is not sufficient to withstand the combined loads.

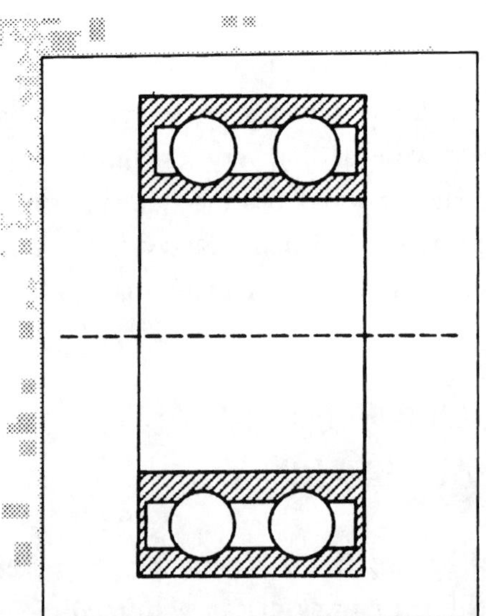

Figure 4.21
Double-row anti-friction ball bearing.

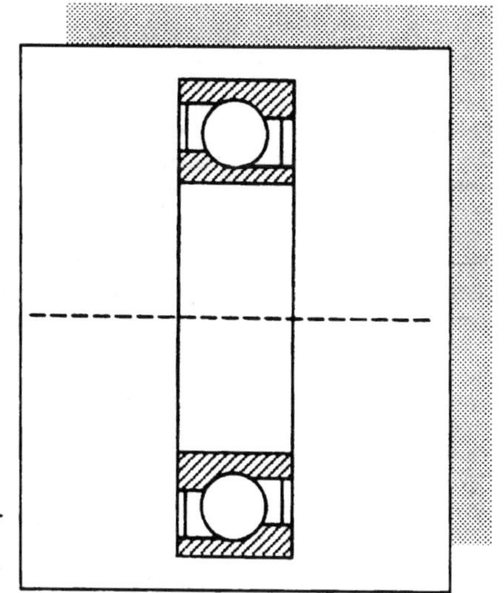

Figure 4.22
Single-row angular contact bearing.

4.7.1.3 ANGULAR CONTACT BEARINGS

The design of 40° angular contact ball bearings operates well under heavy thrust loads. The single-row type (see Figure 4.22) is good for thrust loads in one direction. In most centrifugal pumps, the thrust reverses during start-up, so the thrust bearing must absorb thrust from both directions. Double angular contact bearings (as shown in Figure 4.23) within the same outer face can withstand moderate radial loads.

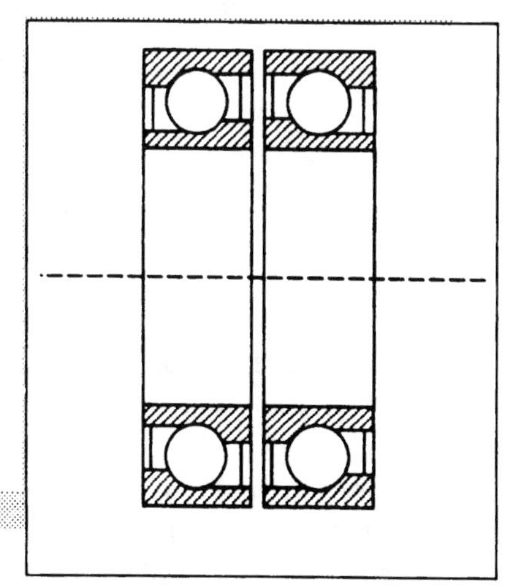

Figure 4.23
Back-to-back angular contact bearing.

TABLE 4.1. Common Anti-Friction Bearings.

Type	Characteristics
Single-row, deep-groove ball bearing	Good radial load and reasonable amount of thrust in either direction. Requires good alignment. High speed.
Angular-contact ball bearing	Heavy thrust in one direction. Moderate radial load. With sides flush ground, can be used in multiple numbers.
Spherical roller bearing	High load capacity, both radial and thrust. Thrust either direction. Self-aligning.
Single-row, tapered roller bearing	Good for heavy one-way thrust or combined load. Predominately thrust load. Requires adjustment of internal clearance.

4.7.1.4 SELF-ALIGNING, SPHERICAL ROLLER BEARINGS

The self-aligning, spherical roller bearing can withstand both heavy radial and thrust loads. The thrust loading can also be in either direction. These bearings find their greatest use on larger shaft sizes for which the selection of suitable ball bearings is limited.

4.7.1.5 SINGLE-ROW, TAPERED ROLLER BEARINGS

The single-row, tapered roller bearing is good for heavy thrust loads in one direction, and, in some cases, it can handle combined loads. This bearing does require adjustment of internal clearances.

IMPORTANT **Note:** *The preceding discussion of bearings is by no means complete, it is intended to provide the reader with some idea of the many different kinds of bearings that are available to meet the* *many different load conditions. While rolling contact bearings are widely used, they are not the only types of bearing used in pumps. Sliding contact bearings, such as sleeve or babbited bearings, have been used in many pumps.*

Table 4.1 provides a summary list of common anti-friction bearings and their characteristics.

4.7.2 BEARING INSTALLATION, MAINTENANCE, AND LUBRICATION

When replacing bearings, the manufacturer's specifications should be followed. If not available, the manufacturer's representative should be consulted.

We pointed out earlier that bearings provide alignment between the moving and non-moving parts. In doing so, there is always a certain amount of friction developed between the bearing and its guides or races. This metal-to-metal contact creates heat that must be transported away from the bearings in order to provide a long running life. To cool the bearing and reduce the amount of friction, oil, grease, or synthetic lubricant is applied. This lubricant must be applied to the moving parts in a thin film. Too much lubrication can increase friction and lead to excessive heat.

When a bearing is installed in a pump or along a shaft, it is placed within a housing that holds the outside of the bearing stationary while providing an area for applying lubrication to the bearing. This housing or seal provides a reservoir for lubrication and encloses the bearing, thus keeping dirt out. There are many different designs for bearing housings, and they vary with the lubrication used.

REFERENCES

AWWA, *Water Transmission and Distribution,* 2nd ed. Denver: American Water Works Association, p. 381, 1996.

Garay, P. N., *Pump Applications Desk Book.* Lilburn, GA: The Fairmont Press, Inc., p. 22, 1990.

Karassik, I. J., Centrifugal Pump Construction, *Pump Handbook,* Karassik, I. J., et al. (eds.). New York: McGraw-Hill, pp. 2-71, 1976.

Renner, D., *Hands-on Water/Wastewater Equipment Maintenance.* Lancaster, PA: Technomic Publishing Co., Inc., p. 184, 1999.

TPC, *Understanding the Operation of Pumps.* Buffalo Grove, IL: TPC Training Systems, p. 141, 1986.

Wahren, U., *Practical Introduction to Pumping Technology.* Houston: Gulf Publishing Company, p. 83, 1997.

Self-Test

4.1 A ______________ casing adds a guiding vane to the fluid passage.

4.2 The impeller does not ______ the water it is pumping, but ________ through the water and throws it.

4.3 A physical separation between the high- and low-pressure sides of a pump is maintained by ______________.

4.4 The ___________ directs water flow into and out of the pump.

4.5 The function of a pump's impeller is to ______________________________.

4.6 The close-coupled pump has no ____________.

4.7 The __________ impeller is used mainly for pumping water containing large solids.

4.8 Packing should be replaced when tightening, the ______________ cannot control leakage.

4.9 The contacting surfaces of the ____________________ are highly polished to ensure proper sealing.

4.10 Load perpendicular to the shaft is called ______________________________.

4.11 Name three types of impellers.

4.12 Identify the impeller type most commonly used in wastewater pumps.

4.13 What is the purpose of the stuffing box/packing assembly?

4.14 Why are bearings important in the operation of a centrifugal pump?

4.15 What is the purpose of the coupling in a pump-motor assembly?

4.16 What is the purpose of the shaft sleeve?

Centrifugal Pump: Operational Procedures

Once a pump has been installed, chances are that it will remain in operation whether or not performance is as expected. Pumps are sometimes purchased oversized, with expected duty to be heavier as system needs increase. A pump which is too powerful for the present system does not last longer. The pump curve will not match the system curve and operation will be off maximum efficiency. On the other hand, a pump operating at design rating when first installed, will not be as efficient as system needs change with time.[23]

TOPICS

Installation
Startup
Normal Operation
Shutdown
Priming
Backflushing

5.1 INTRODUCTION

The proper operation of a centrifugal pump helps to ensure that the pump maintains its peak performance with minimal amounts of downtime and/or unexpected and costly maintenance. However, it is important to note that neither good operational procedures or maintenance programs can ensure the smooth operation of any pump that has not been properly selected for its specific application. For example, if a standard centrifugal pump with cast-iron casing is used to pump gritty abrasive materials, it is

[23]Hauser, B. A., *Practical Hydraulics Handbook,* 2nd ed. Boca Raton, FL: Lewis Publishers, p. 141, 1996.

Key Terms Used in This Chapter

BACKFLUSHING ▶ Using water to flush the pump to remove debris

PRIMING ▶ Filling the pump with water each time it is started

unlikely that the pump casing (or impeller; see Figure 5.1) will have a long operating life no matter how conscientiously the unit is operated and maintained.

This chapter will cover important aspects related to typical operational procedures used in the proper operation of centrifugal pumps including pump installation procedures, pump startup procedures, normal operation, pump shutdown procedures, pump priming procedures, and procedures typically used for backflushing.

Figure 5.1
A centrifugal pump impeller that was used to pump gritty and corrosive solutions.

IMPORTANT

Important Note of Caution: *The normal operation of a centrifugal pump is not complex. In fact, ease of operation remains one of the major advantages of the centrifugal pump. The level of technical skills required to operate a centrifugal pump is much less than that required for the majority of the other classifications of pumps. Nevertheless, the procedures discussed in the following are* generic; *that is, they are typical of centrifugal pump operations used in many water/wastewater operations. Because the operational procedures discussed are generic and/or typical does* not *mean that they are "the"* exact *procedures that should be used at any particular plant for any particular centrifugal pump. You must ensure that, for any specific centrifugal pump operation, the manufacturer's technical manual is the first source of correct operational instructions. If the manufacturer's technical manual is not available, check with the vendor and/or the person in charge of the pump to ensure that the pump is operated in accordance with proper operating requirements.*

5.2 INSTALLATION

Technically, pump installation may not fall within the parameters of general operation, but it can have a significant impact on the routine operation of the pump.

All major reputable pump manufacturers/vendors supply specific installation instructions with their pumps. These include

- pump specifications
- electrical service requirements
- detailed installation instructions
- foundation specifications
- control system specifications

These pump installation instructions (supplied by the manufacturer/vendor) should be included in the plant's equipment records.

5.2.1 INSTALLATION PROCEDURE

If a centrifugal pump is to be purchased and installed by plant personnel, we recommend that the pump purchase requisition include a request for vendor installation procedures. Moreover, according to Wahren,[24] even if the delivered pump package includes installation instructions, the purchaser should consider the following items:

1. Upon receipt of the pump package, the purchaser should inspect the package for any signs of damage during transport. If damage is noted, the pump package should be immediately returned to the manufacturer/vendor. Remember, once you sign for an item (any item), you take ownership and could end up accepting damaged goods—and become stuck with them. The purchaser should also ensure that the package conforms to the bill of lading.
2. All lifting must be per vendor's instructions. Do not use slings around nozzles or other openings not designed for lifting.
3. Remove dirt, grease, and oil from equipment feet, or bottom of skid or baseplate. Ensure that the vendor provided protection from damage and corrosion for these parts during transportation and storage. Again, do not sign for any pump package where damage is apparent.
4. When positioning the pump package for mounting, the use of type 316 stainless steel, 1/4-in. thick leveling shims are preferred.
5. Generally, pump bases with four anchor bolts require a set of shims for each bolt.
6. Pump bases with six or more anchor bolts require U-type shims when positioning the pump package.
7. Wedges should not be used.
8. As a general rule, the installation tolerances are
   - 1/8 in. in any direction of horizontal displacement
   - ± 1/8 in. elevation
   - 1/2° for face alignment in vertical and horizontal plane
9. The pump package should not be leveled or aligned by tightening bolts.
10. The largest flange, usually the suction flange, is the position selected for aligning the equipment.

[24]Wahren, U., *Practical Introduction to Pumping Technology.* Houston: Gulf Publishing Company, pp. 146–147, 1997.

11. If the pump and driver lie on the same baseplate or skid, disconnect the two before leveling the equipment.
12. Disconnect the coupling between the pump and the driver before positioning and leveling the equipment.
13. Make up the coupling after positioning and leveling the equipment, using the vendor's tolerances as a guide. Do not dowel the bases until hot aligning the package, if required.
14. Proceed with grouting of the equipment.
15. Use a dial indicator and reverse indication for parallel and angular coupling alignment. The tolerances must be within ±0.005 in. for both rigid and flexible couplings.
16. Misalignment causes between 50 and 70% of all pump bearing failures, so the purchaser may want to use stricter alignment tolerance and a laser indicator to get more accurate readings.

5.3 STARTUP

As is the case with pump installation, major pump manufacturers supply valuable information on the proper procedures to use when starting a pump for the first time or for situations where the pump is being placed back in service after removal for repair.

Again, it is essential that the pump be thoroughly inspected and placed in service by someone who is familiar with the correct startup procedure. If the pump is not put into service properly, it may be damaged or its operating life may be dramatically shortened. It should be noted that if the pump is improperly installed or started, the damage that occurs might not be covered by the pump warranty.

It is "always" good practice to ensure that any contract for a new pump, and/or new pump installation, also includes provisions for factory-trained service personnel to install, startup, and train plant personnel on proper startup/operational procedures. Unless the plant staff is highly experienced in the area of pump operation and maintenance, this may prove to be the least costly method to ensure proper installation and initial operation.

All technical literature supplied by the manufacturer regarding pump startup should be maintained as part of the plant's permanent records. Moreover, in many circumstances, it is highly beneficial to post written startup procedures at the pump operation station.

For situations where the manufacturer does not provide specific instructions, the following generic procedure may be helpful in starting the centrifugal pump. (Note: Remember, the manufacturer's startup procedures always take precedence over any other procedure.)

5.3.1 STARTUP PROCEDURE

1. Inspect pump bearings and lubricate if needed. (Note: Be careful not to over-lubricate the bearings—too much lubrication can be just as damaging as too little lubrication.)
2. If possible, turn pump shaft by hand to ensure free rotation.
3. Check shaft alignment, and adjust if necessary.
4. Check coupling alignment, and adjust if necessary.
5. Check electrical service to the pump to ensure it is wired correctly and has the appropriate service (i.e., 120, 240, 440 volts, single or three phase, etc.).
6. Check motor heaters, and reset if necessary.
7. Turn motor on, then off—check rotation of the pump shaft. (Proper rotation is normally indicated on the pump case.)
8. Inspect pump control system. Does it activate the motor starter when the liquid level reaches the desired level? Does it deactivate the pump motor when the level reaches the desired cut-off point? Adjust as necessary.
9. Adjust packing by tightening packing gland until nuts are finger-tight.
10. Start seal water flow, and allow a high rate of leakage during startup.
11. Prime pump by filling casing and suction line.
12. Start pump in the manual mode of operation.
13. Monitor pump operation, bearing temperature, delivery rate, and discharge pressure. Record for future reference.

14. Observe operation of control system.
15. Adjust seal water flow to desired flow rate.
16. If all systems are found to be operating as designed, place pump system in "auto" control mode, and set operation clock to record actual operating time.
17. Do not operate the pump with the discharge or suction valve closed except for a very short time, because this will damage the pump.

5.4 NORMAL OPERATION

Normal operation requirements for a centrifugal pump are relatively simple and straightforward. It is a matter of housekeeping, observation, lubrication, and maintenance. Maintenance and lubrication procedures are addressed in Chapters 6, 7, and 8. In general, normal operation consists of the following.

1. Carefully observe the pump operation, paying particular attention to the sounds of the operation, the amount of vibration, and operating temperature of the bearings and motor.
2. Observe and adjust seal water flow rate (approximately 20 drops/minute leakage for conventional packing).
3. Observe control system operation.
4. Clean control sensors.
5. Observe control valves and check valves.
6. Observe the discharge volume or pressure.
7. Record pump parameters and cumulative operating time.
8. Rotate pumps to ensure even wear on available pumps (usually performed at least once per week).

5.5 SHUTDOWN

As with installation, startup, and normal operation, the manufacturer's instruction book also normally includes very specific instructions regarding the shutdown of the pump for either routine maintenance or for extended periods. Again, because the manufacturer is the authority on its pump and has better knowledge of the pump's requirements, these recommendations should be followed whenever possible.

In the rare event that this information is unavailable, the following general procedure may be of assistance to ensure that the pump is not damaged by its removal from service.

1. Place alternate pump in service.
2. Observe alternate pump's operation to ensure continued availability of pumping capacity.
3. Shut pump off.
4. Close intake and discharge control valves.
5. Close seal water valve.
6. Open circuit breaker, lockout/tagout.
7. Flush the pump to remove wastes and debris from casing. Allowing waste to remain in the pump could be detrimental to the pump and to workers. The trapped wastes could produce sufficient gases to rupture the case and/or create flammable or explosive conditions.
8. Perform scheduled maintenance to ensure pump is ready to return to service when necessary.

5.6 PRIMING

Hauser[25] points out that a major deficiency of centrifugal pumps is that the pump chamber must be filled completely with water upon startup for it to function correctly. However, to combat this deficiency, in the majority of the installations, the centrifugal pump is installed to allow gravity flow from the wet well. In this way, the pump fills by gravity, ensuring that the inside of the pump casing and impeller are constantly full (i.e., fully primed). It is essential that the casing and impeller be full of water, or the pump will either not deliver any liquid or, if partially filled, the pump will discharge at a reduced rate.

[25]Hauser, B. A., *Hydraulics for Operators.* Boca Raton, FL: Lewis Publishers, p. 100, 1993.

5.6.1 PRIMING PROCEDURE

As previously mentioned, it is essential that the entire pump interior be completely filled with water and that no air be trapped within the casing. In addition to the reduced capacity associated with lack of a full prime, the pump may also be extremely noisy and may vibrate excessively. If allowed to operate for any length of time in this condition, mechanical damage to the impeller, casing, or shaft may occur. The following is a general priming procedure for the centrifugal pump.

1. Open the vent valve at the top of the casing. (Note: Depending upon the pump location, the next step will vary.)

For pumps located below the liquid level in the wet well,

2. Slowly open the valve on the intake or suction side of the pump.
3. Allow the casing to fill until liquid is leaving the pump casing from the vent valve.
4. Close the vent valve.
5. Start pump.
6. Slowly open discharge valve until fully open.

For pumps located above the level of the liquid in the wet well,

2. Open the pump suction and discharge valves.
3. Slowly and carefully, open the pump's discharge check valve to allow a backflow from the discharge line into the pump casing. (Note: If the pump is part of a multiple pump setup, the discharge line will be under pressure, and extreme care must be used in opening the discharge check valve.)
4. Allow the flow to continue until water is noted at the vent valve. Close the vent valve.
5. Slowly close the discharge check valve. Rapid closure of this valve could permanently damage the check valve as well as perhaps cause serious damage to the associated piping.
6. Close discharge control valve.

7. When ready to start pump, start motor and slowly open the discharge valve.

In a few instances, the pump casing and impeller can't be filled by either of the means described due to the lack of a source of water that can flow into the pump. In these cases, the pump may be manually filled. The procedure in this case is

1. Open the vent valve.
2. Remove plug located near the top of the casing or near the top of the discharge of the pump.
3. Pour water slowly into the pump casing until liquid is noted at the vent valve.
4. Replace plug.
5. Start the pump as described previously.

If manual filling of the pump casing is not feasible or if the pump can't be expected to maintain a prime during off periods, it may prove beneficial to install some form of vacuum system that will allow the operator to withdraw air from the casing. The removal of air will create a vacuum and will draw water from the wet well or supply tank to the pump. Although more expensive than the manual filling of the casing, it is very beneficial for pumps that require repeated priming.

5.7 BACKFLUSHING

In many situations, debris may be trapped within the casing and/or impeller. When this occurs, the pump may not discharge at its rated capacity, and/or it will operate with more noise and vibration. To correct this problem, the pump casing and impeller must be cleaned to remove the debris. There are two methods available to accomplish this:

1. Backflush the pump.
2. Manually remove the debris through the inspection ports or by disassembling the pump casing.

5.7.1 BACKFLUSH PROCEDURE

Note: Check pump operation and pump prime to determine if the cause of the excessive vibration and noise is debris trapped in the casing or impeller.

1. Inspect pump to ensure there is no mechanical cause for vibration.
2. Open vent valve on casing to ensure the pump is fully primed.
3. Shut off pump.
4. Close discharge valve.
5. Slowly open discharge check valve, and hold it open.
6. Turn on second pump that operates on the same discharge line.
7. Slowly open the discharge control valve on the pump requiring backflushing.
8. Allow the backflushing to continue for a few minutes.
9. Slowly close the control valve on the pump being backflushed.
10. Slowly close check valve. Do not allow the valve to "slam" because that could cause serious damage to the valve.
11. Start pump that has been backflushed. Slowly open the discharge control valve.
12. Observe operation. If discharge is still below the normal level and/or the pump is still noisy and vibrating, remove the pump from service for additional maintenance.

IMPORTANT

***Note:** In some instances, plant personnel can utilize methods that will allow the pump to be backwashed even if there is no other pump available on the line to provide the pressure. If the pump is found to clog frequently, it may be possible, under proper supervision by an experienced operator, to use a tap and valve on the pump's discharge line to connect an outside pressure source or auxiliary pump. However, never use a positive-displacement pump without the necessary safety equipment to prevent excessive exposure.*

IMPORTANT

***Cautionary Note:** Never use wastewater to flush a pump. It is important that no cross-connections be created that would allow wastewater to enter the city water supply.*

5.7.2 MANUAL REMOVAL PROCEDURE

1. Follow the procedure outlined in Section 5.5 to isolate and shut down the pump.
2. After checking to ensure the pump's electrical circuit has been opened and properly locked/tagged out, remove inspection ports or hand holes.
3. Using a flashlight or other light source, inspect the pump casing interior.
4. Using gloves or other safety equipment, remove any debris located within the casing.
5. Flush the interior of the casing with service water.
6. If possible, rotate shaft to ensure debris has been removed.
7. Replace inspection ports.
8. Prime the pump.
9. Place the pump back in service.
10. Observe operation. If still vibrating or discharging at a reduced capacity, remove from service, and lockout/tagout before disassembling to determine cause.

REFERENCES

Hauser, B. A., *Hydraulics for Operators.* Boca Raton, FL: Lewis Publishers, 1993.

Hauser, B. A., *Practical Hydraulics Handbook,* 2nd ed. Boca Raton, FL: Lewis Publishers, 1996.

Wahren, U., *Practical Introduction to Pumping Technology.* Houston, TX: Gulf Publishing Company, pp. 146–147, 1997.

Self-Test

5.1 Ideally, who should perform the initial startup of a new pump?

5.2 List four items that should be checked during the initial startup of a new pump.

5.3 Daily operation of a pump should include inspection of the pump. List four items that should be checked.

5.4 Describe one procedure for priming a centrifugal pump.

5.5 Describe a procedure for backflushing a centrifugal pump.

Centrifugal Pump: Maintenance Procedures

Maintenance is a part of everyday life, although little thought is given to some of the more routine "chores," such as car and home repairs, lawn and shrubbery care, painting, and many other items. However, even though these chores are considered the preservation of property or equipment, they are a form of maintenance. And when you look at the bottom line, maintenance really is the preservation of property.[26]

TOPICS

Pump and Motor Lubrication
Packing and Seal Replacement
Pump and Motor Bearing Inspection
Shaft and Coupling Alignment
Removal of Obstructions

6.1 INTRODUCTION

If we accept Renner's opening statement as valid, it logically follows that the family home, car, lawn mower, swimming pool, HVAC system, and other home appliances all need routine maintenance. As with most of the possessions we own, every centrifugal pump also requires a certain amount of routine maintenance. Whether it is a daily inspection and adjustment of the packing gland or a yearly inspection and adjustment of the shaft and coupling alignment, a centrifugal pump requires some maintenance to provide reliable service throughout its normal operational life.

[26]Renner, D., *Hands-on Water/Wastewater Equipment Maintenance.* Lancaster, PA: Technomic Publishing Co., Inc., p. 1, 1999.

Key Terms Used in This Chapter

ROUTINE MAINTENANCE ▶ Those operations that do not require the use of machine shop equipment and normally can be performed by one person

6.2 PUMP AND MOTOR LUBRICATION

Lubrication (i.e., using a controlled amount of grease or oil lubricant) is probably the most important and most frequently performed of all maintenance functions. This is the case, of course, because lubrication is needed to overcome friction, reduce heat buildup, and extend the life of motor and pump bearings. Therefore, the need for proper lubrication techniques can't be over-emphasized. It is important to note that both over- or under-lubrication can damage the motor and/or pump and create excessive downtime for pumping equipment. Accordingly, the maintenance operator in charge of lubricating the motors and pumps should be familiar with the manufacturer's recommendations for lubricants and lubrication procedures. The recommendations are supplied with the pump and/or motor and usually vary from pump to pump.

Note: *Keep in mind that the manufacturer's recommendations normally take precedence over any other guidelines specified for pump maintenance.*

IMPORTANT

Note: *In Chapter 8, we provide general or idealized lubrication procedures. These procedures are "general" in form—they are intended to be reviewed and modified to meet the specific requirements for each individual application. The information provided is based on experience and various reference sources. It is intended to point out the important features or steps that are needed for a successful lubrication program. However, under all circumstances, the manufacturer's technical manual should always be consulted to ensure proper lubrication steps are followed.*

6.3 PACKING AND SEAL REPLACEMENT

The stuffing box should be inspected each day the pump is in operation. During the inspection, the amount of leakage should be noted. If the gland is adjusted properly, a leakage rate of 20 to 60 drops of seal water per minute is normal. Inadequate or excessive amounts of leakage are signs of possible trouble.

If the leakage is below 20 drops per minute or there is no leakage, the operator should

1. Gradually loosen the packing gland nuts.
2. Observe the leakage rate after loosening the gland.
3. If the rate increases, assume that the gland had been over-tightened. The operator should allow the stuffing box to cool and then shut the pump down. After locking/tagging the pump out, the gland and packing should be removed. With the packing out, the shaft or sleeve can be inspected for excessive wear. If the shaft or sleeve is not damaged, the pump should be repacked following instructions provided in Section 6.3.1.
4. If after loosening the gland, no leakage is observed, the pump should be shut down and locked/tagged out. At this time, the seal water supply system should be investigated following the manufacturer's troubleshooting procedures. Once the seal water problem is corrected, but before the pump is put back in service, the gland and packing should be removed, and all components should be inspected. If there is excessive scoring of the shaft or sleeve, this should be repaired before the pump is repacked and put in service.
5. If excessive leakage is occurring, the gland should be tightened following the procedure outlined for repacking the pump. If the amount of leakage can't be controlled, the pump should be repacked.

6.3.1 PACKING PROCEDURE

The following presents a common packing procedure.

IMPORTANT **Note:** *Only experienced personnel should attempt the repacking and adjusting of stuffing boxes. If those with little experience must do the job, they should be cautioned against placing too much pres-*

sure on the gland. They should be informed that excessive leakage is not as damaging as too little.

1. Never try to add one or two rings to the old packing. It is false economy. Remove the old packing completely, using a packing puller if available, and clean the box thoroughly. Inspect the sleeve to make sure it is in acceptable condition. Putting new packing against a rough or badly worn sleeve will not give satisfactory service.
2. Make sure that the new packing is a proper type for the liquid, operating pressure, and temperature. Unless the packing is die-molded and in sets, make sure that each ring is cut square on a model of correct size.
3. Insert each ring of packing separately, pushing it squarely into the box and firmly seating it by using the packing gland or two half rings of proper length. Successive rings of packing should be staggered so the joints are 120° or 180° apart.
4. When a lantern ring is involved, make sure it is installed between the correct two rings of packing so it will be located properly for the sealing liquid supply when the box is fully packed and adjusted.
5. After all the required rings of packing have been inserted, install the gland, and tighten the gland nuts firmly by hand. In doing so, make sure that the gland enters the stuffing box squarely and without cracking, so the outer edge of the packing is compressed uniformly.
6. After this first tightening of the gland, back off the nuts until they are merely finger-tight. When ready, start the pump, keeping the gland loose so there is excessive leakage initially. Periodically tighten up slightly (1/8 of a turn) and evenly on the gland nuts so the leakage is reduced to normal after several hours. Do not attempt to reduce the leakage too much. It must be a steady stream sufficient to carry away the heat generated by the packing friction.

6.3.2 MECHANICAL SEAL INSTALLATION PROCEDURE

The routine maintenance for mechanical seals involves inspecting the seals daily, ensuring that the seal water is flowing, and replacing the seal when it no longer prevents leakage. Those responsible for maintenance of pumps employing mechanical seals should carefully read the seal manufacturer's instructions for the operation and maintenance of the seal. Because of the wide variation in seals being used, it is difficult to describe a step-by-step replacement procedure similar to the one for packing systems discussed in Section 6.3.1.

IMPORTANT

Note: *To obtain satisfactory service and long life out of a mechanical seal, a small amount of seal water (a drop or two every few minutes) is required at all times during operating periods. A seal that runs "dry" will fail rapidly.*

The outline that follows provides a few general steps that apply to most seal replacements or installations. Again, the manufacturer's technical manual (or literature) provided with the mechanical seal is the best source of instructions and should always be used when available.

1. Shut the pump down, and lockout/tagout the system.
2. Close the suction and discharge valves, and remove the drain plug.
3. Dismantle the pump, and inspect the shaft or shaft sleeve. If a mechanical seal is being installed to replace conventional packing, the shaft sleeve needs to be replaced. If the mechanical seal is being replaced with another seal, the shaft or sleeve should be cleaned with emery cloth.
4. Clean the shaft and/or sleeve to remove any filings. If the shaft or sleeve is pitted or corroded, it should be replaced.
5. Check the shaft for end play and runout. End play can't exceed 0.005", and the runout should be less than 0.001" per inch of shaft diameter. If shaft end play or runout is excessive, the shaft bearings or the shaft should be replaced.
6. Spray or brush layout bluing on the shaft around the area of the seal housing.
7. Re-install the seal housing, and mark the location of the top of the housing on the shaft; remove the housing.
8. Using manufacturer's specifications, mark the location of the rotating element on the shaft.
9. Before installing the rotating element, check the edge of the shaft for burrs that could cut the O-ring secondary seal.

10. Remove the seal from its container; care must be taken not to damage the primary sealing faces.
11. Position the rotating element on the shaft at the marked location; fasten it down temporarily.
12. Place the stationary element into the seal housing, and install the housing on the pump.
13. Using the feeler gauge, adjust the rotating element to establish the proper clearance; fasten the element in place.
14. Reassemble the pump, and put it back in service; check seal operation.

6.4 PUMP AND MOTOR BEARING INSPECTION

Normally, proper application, proper lubrication, and the use of antifriction bearings on centrifugal pumps and their drive units will ensure they have a long life. Thus, as part of the normal daily operating routine, pump and motor bearing maintenance is limited to a hand check of bearing temperature. (Note: Ideally, a digital read-out, portable thermometer should be used to check bearing temperatures. However, a piece of clay and standard bulb thermometer can also be used.)

IMPORTANT **Note:** *Daily hand checks of bearing temperature are a crude but often effective method of determining bearing operating conditions.*

Beyond the normal daily operating routine, bearing temperatures are routinely checked using a thermometer every month. The reading obtained should be compared with the previous readings. Normal operating temperatures are approximately 180°F. However, bearing temperatures do vary with each installation. The actual bearing running temperature is not that important. What is important is spotting temperature increases above normal operating levels before serious damage occurs in the unit.

Rising bearing temperatures are an indication of increased friction within the bearing and are a sign of problems (e.g., over-lubrication, improper pump and motor alignment, and under-lubrication). When elevated temperatures are discovered, the cause of the problem should be investigated immediately.

6.5 SHAFT AND COUPLING ALIGNMENT

Shaft and coupling alignment should be inspected at least every six months. However, if there are any signs of problems because of misalignment, an inspection should be performed immediately. Correct alignment of the pump and driver and any intermediate shafting and couplings is very important to trouble-free operation. Keep in mind that a flexible coupling will not compensate for all misalignment. Noisy pump operation, reduced bearing life, excessive coupling wear, and waste of power may result from faulty alignment.

The procedures that should be followed when checking and/or installing a coupling are not always the same. The manufacturer determines the proper alignment procedure for any given coupling. These alignment instructions should be included in the operating manuals that are furnished with any purchased machinery. If the alignment instructions are missing, there are a few basic procedures that can be used.

6.5.1 ALIGNMENT PROCEDURE

1. On new installations, the pump is leveled, and a preliminary alignment is made before grouting the base plate to its foundation. After the grout has set for not less than 48 hours and foundation bolts are tightened, the driver is now ready for alignment.
2. Start the alignment process with a check and/or correction of the angular alignment. The coupling gap should be checked with a feeler gauge or a coupling alignment indicator gauge. The check should be done in both the horizontal and vertical plane. Use shim stock no smaller than the motor feet to make any necessary corrections in angular alignment.

3. After proper angular alignment has been established, any parallel misalignment needs to be corrected. Using a straight edge, align the drive and pump unit couplings so they line up with each other on all sides. Once again, use shims to adjust the drive unit and establish proper alignment.
4. A better method of checking angular or parallel alignment is with the use of a dial indicator (see Figure 6.1).
5. Remember: Re-check each alignment after making any adjustment.

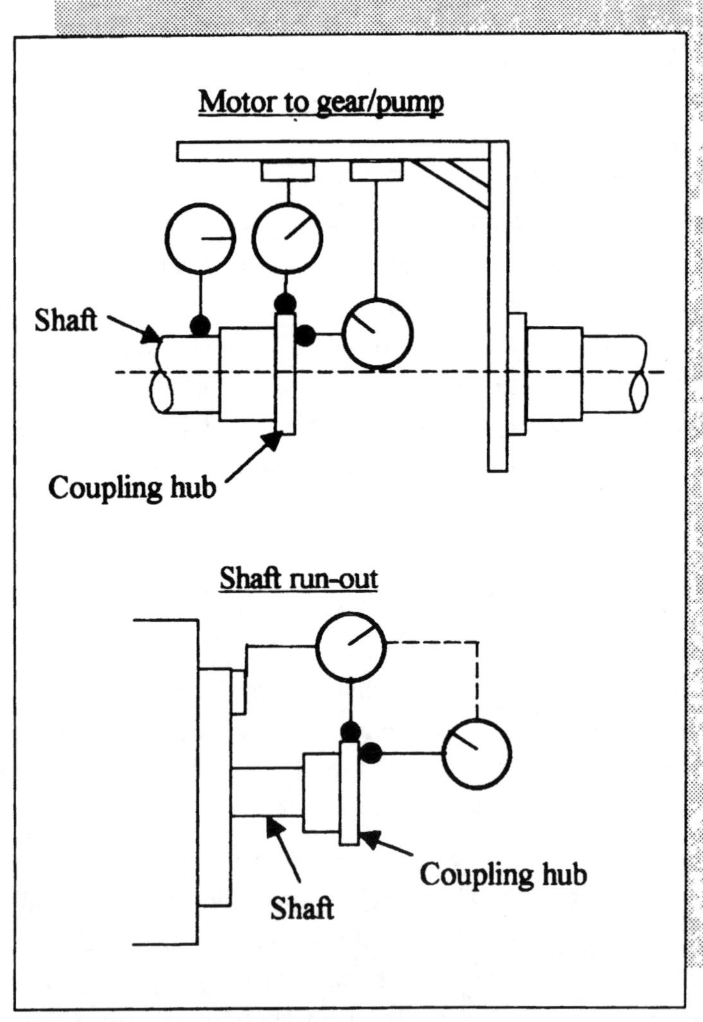

Figure 6.1
The use of a dial indicator to align pump.

6.6 REMOVAL OF OBSTRUCTIONS

It is often necessary to open the pump to remove obstructions that have lodged in or around the impeller and/or the volute. This procedure is rather basic; however, certain precautions should always be taken.

1. Always "shut down" and lockout/tagout the system.
2. Check the suction and discharge valves to ensure they are closed.
3. Remove vent or drain plug. This allows the pressure and water to escape from the volute.
4. After the flow has stopped, open the pump.
   - Solid case pumps are usually equipped with removable inspection plates. After removing the plate, the internal portion of the pump is accessible.
   - Split case pumps do not have inspection plates; however, half the casing can be easily removed for internal inspection.
5. Remove the obstruction, and inspect the inside of the pump.
6. Make any necessary repairs, and then put the pump back together.
7. Prime the pump, and put it into operation. Vent any entrapped air from the volute. This is done by slowly opening the vent plug on the volute and allowing the air to escape. Caution should be taken when performing this operation because the inside of the volute is under pressure.

REFERENCE

Renner, D., *Hands-on Water/Wastewater Equipment Maintenance.* Lancaster, PA: Technomic Publishing Co., Inc., 1999.

Self-Test

6.1 The major maintenance problems in centrifugal pumps occur in the ____________.

6.2 Describe the procedure for checking the alignment of the motor and pump shafts.

6.3 Describe the procedure for repacking the centrifugal pump stuffing box.

6.4 Describe the procedure for replacing a mechanical seal.

6.5 What determines the amount of maintenance that can be performed by plant maintenance personnel?

Centrifugal Pump: Preventive Maintenance

We can categorize maintenance as either reactive or preventive. Reactive maintenance occurs when equipment fails unexpectedly. Preventive maintenance occurs according to an organized method to address potential problems.

Reactive maintenance leads to a plant in a constant state of "putting out fires." The fire is, of course, the equipment breakdown. Putting the fire out is restoring the equipment to fully operational status. The basic problem with reactive maintenance is rather obvious. To extend the analogy, preventing a fire is always better than fighting a fire.

Effective preventive maintenance prevents a lot of fires—unscheduled equipment shutdowns.

Which do you prefer?

TOPICS

Daily Maintenance
Weekly Maintenance
Monthly Maintenance
Quarterly Maintenance
Semi-Annual Maintenance

7.1 INTRODUCTION

The centrifugal pump (and several other types of pumps) used in water/wastewater operations is considered to be a relatively low maintenance hydraulic machine. However, there is a direct relationship between the service obtained from the pump and the preventive maintenance (i.e., the scheduling and performance of regular maintenance functions from inspections through overhaul or replacement; typically consists of planned and/or routine maintenance) that the pump receives.

Key Terms Used in This Chapter

PLANNED MAINTENANCE	The periodic scheduled removal of equipment from service for the disassembly and inspection of the internal working parts
PREVENTIVE MAINTENANCE	The scheduling and performance of regular maintenance functions from inspections through overhaul or replacement
ROUTINE MAINTENANCE	The daily general inspections and tests, including regular lubrication, that are performed as part of the regular operation of the plant

In many instances, a pump may be capable of providing excellent service for an extended period with little or no attention. However, if there is no planned maintenance (i.e., the periodic scheduled removal of equipment from service for the disassembly and inspection of the internal working parts) or routine maintenance (i.e., daily general inspection and testing, including regular lubrication, performed as a part of the regular operation of the plant), the end result will be a shortened service life and eventual decay of the equipment.

Again, the manufacturer's technical manual is always the best source of specific maintenance information. Based on service history, case study, and controlled testing, the manufacturer typically outlines those items that require attention and the frequency for this attention. Accordingly, whenever possible, the manufacturer's literature should be consulted before developing a schedule of maintenance for any pump. If the manufacturer's technical manual is not available, the manufacturer's representative should be consulted for assistance in developing a preventive maintenance schedule.

In cases where neither the manufacturer's technical manual or manufacturer's representative is available to assist you, the general preventive maintenance schedule, based on periodicity (see Figure 7.1), may be of assistance. (**Note:** It is important to point out that the idealized schedules, based on periodicity, presented in the following sections may be either too simple or too complex for any particular application. They are presented

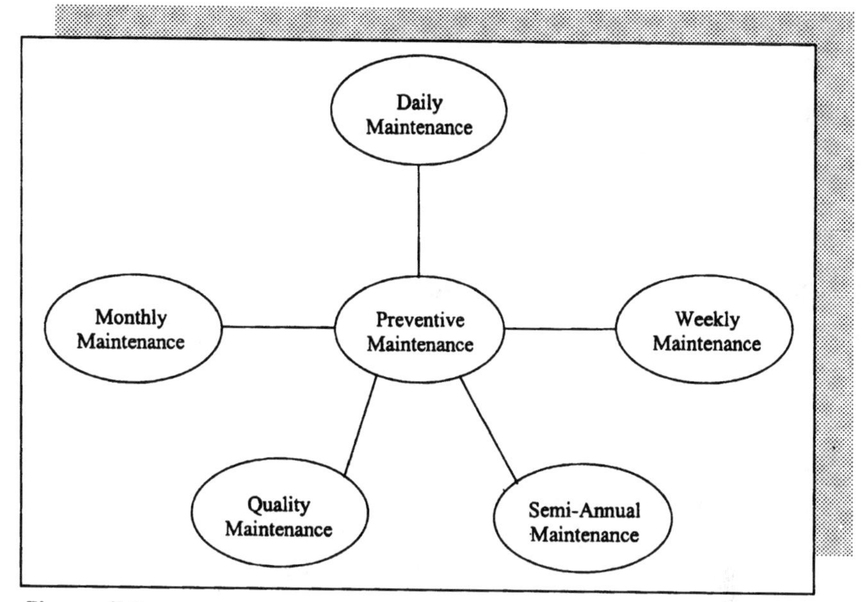

Figure 7.1
Periodicity of preventive maintenance.

as a starting point for development of an overall preventive maintenance schedule. Each pumping situation may require this schedule to be modified or tailored to reflect on-site personnel experience and operating conditions.)

Important Point: *The term "preventive maintenance" should not be assumed to be synonymous with another term that is commonly used today:* predictive maintenance. *Predictive maintenance is an advanced form of preventive maintenance. Predictive maintenance strategies seek to apply maintenance techniques* ***only when needed,*** *based on information such as vibration and thermographic and lubricant condition analyses.*[27]

7.2 DAILY MAINTENANCE

For centrifugal pumps, the following observations, determinations, and/or adjustments should be made on a daily basis.

[27]WERF, *Improving Wastewater Treatment Plant Operations Efficiency and Effectiveness.* Alexandria, VA: Water Environment Research Foundation, pp. 3–10, 1999.

1. Visually observe pump operation.
2. Determine approximate operating temperatures for motor and bearings (by touch).
3. Adjust seal water flow rate.
4. Adjust packing gland.
5. Refill or adjust oil or grease sealed packing gland.
6. Clean sensor unit for pump control.
7. Check foundation bolts.
8. Observe motor operation.

7.3 WEEKLY MAINTENANCE

For centrifugal pumps, the following observations, determinations, and/or adjustments should be made on a weekly basis.

1. Alternate operation. If two or more pumps of equal capacity are available, place second pump in operation.
2. Clean pump. Lockout/tagout pump system, then removal all debris from inside the casing.
3. Check packing assembly. If leaking excessively after tightening, remove packing and inspect shaft and/or shaft sleeve. If shaft or sleeve is badly grooved or extremely rough, make necessary arrangement to have repairs made.
4. Check operation of control system: is system starting the pump at the desired level and stopping the pump at the desired point? If not, clean and adjust.
5. Inspect motor for indications of overload, burnt insulation, melted solder, etc. If any of these conditions are noted, have a qualified electrician inspect the motor.

7.4 MONTHLY MAINTENANCE

For centrifugal pumps, the following observations, determinations, and/or adjustments should be made on a monthly basis.

1. Adjust packing (if necessary).
2. Check motor ventilation screens.

7.5 QUARTERLY MAINTENANCE

For centrifugal pumps, the following observations, determinations, and/or adjustments should be made on a quarterly basis.

1. Inspect and lubricate pump bearings. Drain lubricant, and clean bearing with a solvent. Flush bearing housing, and clean all moving parts. Inspect bearing seals and bearings for wear. Reassemble bearings, and fill with the specified amount of the correct lubricant.
2. Check bearing temperature with a thermometer.

7.6 SEMI-ANNUAL MAINTENANCE

For centrifugal pumps, the following observations, determinations, and/or adjustments should be made on a semi-annual basis.

1. Check pump-motor shaft alignment.
2. Perform a complete inspection and servicing of pump.
   - determine pumping capacity
   - determine pumping efficiency
   - inspect wear rings
   - inspect check valves and control valves
   - clean scale and debris from pump casing
   - inspect impeller condition, replace if worn
   - inspect pump shaft and shaft sleeve, replace as required

REFERENCE

WERF, *Improving Wastewater Treatment Plant Operations Efficiency and Effectiveness,* Project 97-CTS-1. Alexandria, VA: Water Environment Research Foundation, 1999.

Self-Test

7.1 Briefly explain why preventive maintenance is an essential part of the maintenance program for a centrifugal pump?

Centrifugal Pump: Lubrication

TOPICS

Purpose of Lubrication
Lubrication Requirements
Lubrication Procedures

Just like car maintenance, routine pump maintenance is important to preserve service life. Therefore, ease of maintenance is a key consideration. The purchaser should ask: Does the pump installation simplify all maintenance concerns, such as access to the pump, and does it provide adequate sensors, instrumentation, or telemetry to indicate problems? What are the potential confined space concerns and necessities.[28]

8.1 INTRODUCTION

A centrifugal pump is made up of several parts that are designed to move at extremely high speeds. Lubrication of these moving parts is probably the most important and most frequently performed of all maintenance functions. Lubrication of centrifugal pump moving parts works to reduce friction and improve the efficiency of the pump. It is essential that these parts be lubricated. However, it is important to note that, in almost every case, too much lubrication can be as damaging as too little.

As is the case with other important centrifugal pump information, the manufacturer normally supplies specific lubrication information with each pump. This information is based upon the manufacturer's knowledge of the equipment and is determined by controlled testing and actual field experience.

This chapter explains the purpose of lubrication, lubrication requirements, and lubrication procedures.

[28]Wood, M., Weis, F., and Mowen, J., Centrifugal Considerations. *Operations Forum,* 16, 5, 12–14, May 1999.

Key Terms Used in This Chapter

LUBRICANT ▶ An oily substance, such as oil and/or grease, that reduces friction, heat, and wear when applied as a surface coating to moving parts.

8.2 PURPOSE OF LUBRICATION

Concerning centrifugal pumps, this section discusses the benefits of lubrication illustrated in Figure 8.1.

8.2.1 SEPARATES SURFACES

To reduce friction between moving parts, a thin film of lubricant must separate them. All lubricants, no matter their type, have the ability to separate surfaces. The lubricant's formulation determines the degree

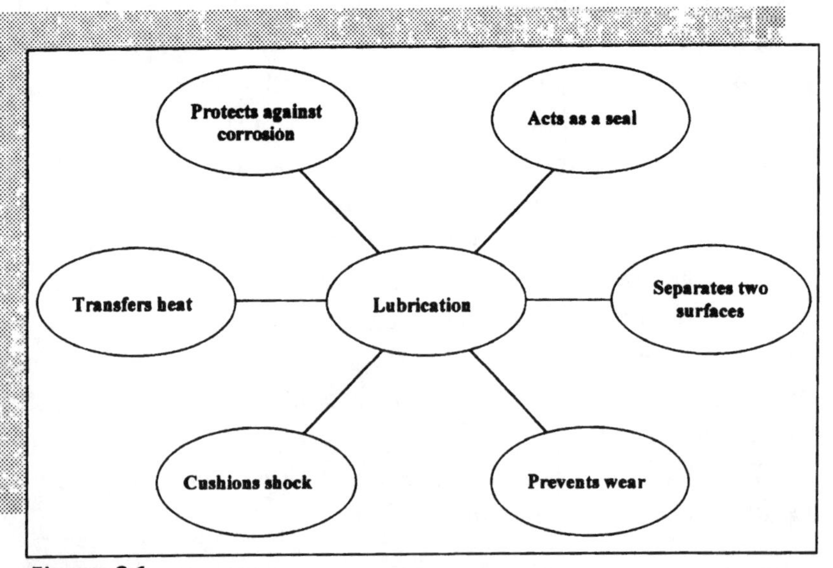

Figure 8.1
Benefits of lubrication.

(amount) of separation between the parts. Lubricants can be grease, oil, or compressed air or other gas.

8.2.2 PREVENTS WEAR

When a lubricant's film layer separates moving parts, it also reduces the amount of wear that takes place as the surfaces rub against one another. The equation is quite simple: Reduced Wear = Longer Life Expectancy. Moreover, as wear does occur (because of the speed of the machine and environmental factors), the lubricant film limits the amount of clearance that occurs as the wear takes place.

8.2.3 CUSHIONS SHOCK

Lubricants provide a cushion between moving parts that dampens shock. The degree of shock dampening provided depends on the lubricant's characteristics and the surfaces it separates. For example, a heavy lubricant, such as grease, is better suited to separating the surfaces and absorbing the shock. The downside of using heavy lubricants, however, is that they do not flow as easily as liquids. Accordingly, when the moving parts are subjected to high speeds, an oil bath-type lubrication is required.

8.2.4 TRANSFERS HEAT

Many maintenance operators have the misconception that lubricants function to absorb heat and, thus, protect machinery. Lubricants do absorb a limited amount of heat, but in fact, they actually work to transfer the heat. If lubricants are subjected to too much heat, they break down. This is the reason that many lubricating systems are water-cooled. Water is an excellent cooling agent.

8.2.5 PROTECTS AGAINST CORROSION

One of the major enemies of any metal surface is corrosion. Once the metal surface of most parts is exposed to the elements, corrosion is the result. However, lubricants coat surfaces and, thereby, work to reduce corrosion. Moreover, most lubricants are formulated with corrosion inhibitors and other chemical additives that improve their film consistency and limit corrosion from occurring.

8.2.6 ACTS AS A SEAL

Depending on the consistency and application of the lubricant, lubrication can provide an excellent protective seal from contamination and moisture.

8.3 LUBRICATION REQUIREMENTS

Whenever lubrication is used, several basic requirements should be kept in mind.

1. The lubricant should be of the type specified by the manufacturer. If a different brand of lubricant is used, it should be specified as equivalent to the recommended brand.
2. Only high-quality lubricants should be used.
3. Lubricants should be added in the amounts and methods specified by the manufacturer. In certain cases, the lubricant is added with the equipment operating; in other cases, the equipment should be idle. Failure to observe the manufacturer's recommendations can result in damage to the bearings or lubrication seals.
4. The lubricant must reach the desired point of application to be effective. If the grease fitting is plugged or, in the case of lubricant lines for bearings on extended shafts, the line is broken or disconnected,

the lubricant will do no good. It is best to inspect lubrication systems frequently to ensure they are working properly.

5. Lubrication responsibility must be assigned to a specific employee to be effective. If left to anyone to accomplish, it is likely that the lubrication will be given a low priority and may not be done as scheduled. Moreover, letting lubrication be accomplished by just "anyone" may be a recipe for pump failure. The point is, the person assigned to perform the lubrication should know how to do it properly.
6. Lubrication systems must be accessible. If it is difficult or hazardous to reach a grease fitting or an oil cup, it is unlikely that anyone is going to be eager to perform the lubrication.
7. In many cases, lubrication may require special equipment or tools. If lubrication is to be performed, these must be readily available.

8.4 LUBRICATION PROCEDURES

Specific lubrication schedules and procedures will vary from one manufacturer to another. In many cases, plant maintenance operators have found it necessary to modify the manufacturer's recommendations based upon their own experience and the specific application of the pump. For this reason, the procedures provided in this handbook are general, idealized procedures that must be reviewed and modified to meet the specific requirements of each individual application.

Important Point: *Proper lubrication means not only regular lubrication, but use of the proper lubricant in the proper amount.*

8.4.1 MOTOR BEARING LUBRICATION

Bearings are a primary part of a motor. In order to operate correctly, bearings must be lubricated with the proper kind of lubricating material in the correct amount. This section provides an idealized motor bearing lubrication procedure.

Note: *We can't overstate the need to follow manufacturer's instructions, if available, on lubricating motor bearings.*

Important Point: *Proper lubrication of motor bearings is extremely important. Without proper lubrication, bearings will overheat, rust, corrode, and eventually cause the shaft to seize and stop. If a motor has been shut down for a long period without proper lubrication, the bearings can become so rusted that the motor will be unable to turn the pump shaft.*

Procedure:

1. Determine whether the motor bearings require lubrication. Many newer motor bearings are composed of metals that are impregnated with lubricant. These bearings do not normally require any additional lubrication. If this type of bearing is used, normally, there are no lubrication fittings located on the bearing assembly.
2. Remove relief plug from bearing assembly (normally located on the opposite side from the grease fitting).
3. Remove any hardened grease from the relief plug, and clean the grease fitting.
4. With the motor running, add four to five strokes of the grease gun to the grease fitting. If the bearing does not have a relief plug, then the new grease should be added very slowly to prevent damage to the bearing's seals.
5. Allow the fresh grease to be heated by running the motor for 5–10 minutes. Leave the relief plug out during this time so that excess grease may drain out of the bearing assembly.
6. Replace the relief plug, and note the date of the lubrication in the maintenance records.

8.4.2 PUMP BEARING LUBRICATION

Pump bearings may be grease or oil lubricated depending on the manufacturer's or supplier's preference. Some special applications use synthetic lubricants. Horizontally shafted pumps may use either oil or grease lubrication while the vertically shafted pump will normally use grease lubrication because of the difficulties that arise when trying to provide a seal that will prevent the loss of the lubricant. Systems to prevent loss of grease are more readily available and are more efficient than those for oil. Although it is more difficult to use oil-lubricated bearings on vertically shafted pumps, there are many cases where they are used.

Important Point: *The type of lubricant used depends upon the pump application and the manufacturer's instructions. If, for example, the pump is operated outdoors in changing temperatures, oil will perform better than grease because its lubricating qualities are not affected by changing temperatures.*

8.4.2.1 OIL LUBRICATION OF PUMP BEARINGS

Oil lubrication of pump bearings is widely used for light-to-moderate, high-speed, horizontally shafted pumps. The design of the oil lubrication system allows oil to be sprayed over the bearing through the use of a slinger ring (see Figure 8.2) or by the movement of the bearing through the oil reservoir, or, in some cases, by the use of an external system that sprays the oil over the bearing.

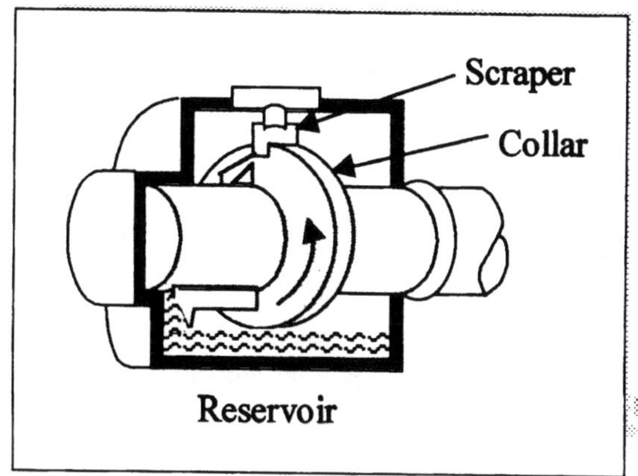

Figure 8.2
Slinger ring oiler.

IMPORTANT **Important Point:** *It is essential that the bearing be coated with a very thin film of oil to provide lubrication without restricting the movement of the bearing. This would increase the operating temperature of the bearing and decrease its operating life.*

8.4.2.1.1 OIL SELECTION

The oils used for bearing lubrication should be filtered, non-detergent mineral oils of grades SAE 10 or 20 without corrosive or abrasive components. Animal or vegetable oils should be avoided, because of the acids formed as these oils breakdown. The oils selected for use must be capable of providing adequate lubrication at startup and at the normal operating temperature of the bearing. If the manufacturer's recommendations are not available, the operating temperature of the bearing and the operating conditions should be discussed with a lubrication specialist to determine the best oil for use.

8.4.2.1.2 OILING PROCEDURE

The normal procedure for oiling pump bearings is a relatively simple matter of draining the oil and refilling the bearing lubrication reservoir with fresh oil.

1. If there is indication that the bearing has been contaminated, it may be necessary to flush the bearing with a suitable solvent. After flushing with solvent, the bearing should be flushed with clean oil to remove excess solvent.
2. The oil reservoir should then be filled to the appropriate level with fresh oil.
3. In some cases, the bearing receives oil lubrication from an external source, such as a reservoir or an automatic oiler. In these cases, daily operation must include checking the delivery of the lubrication to the bearing and refilling the oil reservoir.

8.4.2.2 GREASE LUBRICATION OF PUMP BEARINGS

Grease lubrication of pump bearings is widely used for heavy loads at low to moderate shaft speeds and for vertically shafted pumps where it would be difficult to maintain the necessary reservoir of oil for the bearing. The use of grease allows the lubricant to be placed where it is needed without the need for sophisticated seals to prevent leakage.

IMPORTANT

Important Point: *As a general rule, pumps should be greased about every three months.*

8.4.2.2.1 GREASE SELECTION

Grease lubricants for pump bearings are normally designed to provide oil lubricant in a soap-like base, which allows the oil to liquefy near the surface of the moving parts to provide a light film of lubricant.

IMPORTANT ***Important Point:*** *When grease is used in a pump application, check its grade and consistency to make sure that it is the type specified by the pump manufacturer.*

Again, temperature is a major consideration in selection of the grease; that is, if the pump operates in an environment with a high ambient temperature, a grease with a high melting point should be selected. Major strides have been made in developing multi-purpose grease that can be used over wide temperature ranges and applications. However, many bearing manufacturers still recommend the use of specific soap greases because of the highly dependable nature of these lubricants and the long history of dependable use.

The two most widely used forms of soap-based lubricants are the lime soap-based and soda-based greases. It is very important that the correct base be selected because the base is a critical factor in the degree of lubrication obtained at any specific operating temperature. If the lubricant is too soft, it will flow into the open spaces between the moving parts. If these spaces become filled, the resistance to movement (friction) will cause excessive wear. On the other hand, if the grease is too stiff, the grease will freeze the moving parts of the bearing, making the startup more difficult and possibly causing excessive wear during startup.

8.4.2.2.2 GREASING PROCEDURE

The greasing procedure for pump bearings follows the same basic procedure outlined for greasing motor bearings. Depending upon the severity of use and operating conditions, the lubrication schedule may be as frequent as every 1–3 months (continuous operation) or as little as every 6–9 months for pumps that are not operated frequently. It should be noted that under some conditions, it may be necessary to lubricate much more frequently. Manufacturer's recommendations and experience must be used to determine the best lubrication schedule.

Important Point: *Like excessive oiling, excessive greasing is just as detrimental as too little lubrication and can cause much damage. In addition to generating heat, an excessive amount of grease can rupture lubricant seals. Ruptured seals can allow contaminants to enter the bearing, thus causing bearing failure.*

REFERENCE

Wood, M., Weis, F. and Mowen, J., Centrifugal Considerations. *Operations Forum,* 16, 5, 12–14, May 1999.

Self-Test

8.1 As a general rule, pumps should be greased about every ________.

8.2 The most important part of bearing maintenance is ______________.

8.3 Before changing oil, it is a good practice to ____________________.

8.4 Pump bearings support the ___________ and reduce the amount of ______________ between the shaft and pump frame.

8.5 Grease is used as a lubricant for ___________ loads at low to moderate shaft speeds.

8.6 What kind of lubricant is commonly used for light to moderate duty in high-speed pumps?

8.7 Why is lubrication of pump bearings an important part of any maintenance program?

8.8 What types of lubrication are most frequently used for bearing lubrication?

8.9 What is the purpose of the grease plug located on the opposite side of the bearing assembly from the grease fitting?

8.10 Describe the procedure for lubrication of a bearing.

Centrifugal Pump: Troubleshooting

9

Centrifugal Pump: Troubleshooting

In the next five years, the population of Asian cities will swell by 250 million people, a number equal to the entire U.S. population. Pumps will bring in the required drinking water, remove the wastewater, and also play an important role in producing the new products needed by these new citizens.[29]

TOPICS

The Troubleshooter
Troubleshooting: What Is It?
Goals of Troubleshooting
The Troubleshooting Process
Centrifugal Pump

9.1 INTRODUCTION

Throughout this discussion of centrifugal pumps, it has been pointed out repeatedly that they are capable of providing reliable and trouble-free operation. However, as with any other piece of mechanical equipment, problems do develop. Whether the problem is the result of poor design or installation techniques or the lack of routine maintenance, time and money are lost each time a pump goes down.

This chapter discusses the troubleshooting procedures for 13 common symptoms related to centrifugal pump problems. In addition, the information provided is designed to assist the maintenance operator determine the possible causes and remedies for centrifugal pump problems. The best source of troubleshooting information is the pump manufacturer. The troubleshooting suggestions and the charts provided by the manufacturer often make it possible for an operator to identify a problem and correct it on his/her own.

[29]Worldwide Pump Market Expected to Reach $22 Billion by 2003. *U.S. Water News,* 17, 2, February 2000.

Key Terms Used in This Chapter

TROUBLESHOOTING ▸ The art of problem solving

9.2 THE TROUBLESHOOTER[30]

Few would argue that in order to paint a masterpiece along the lines of the great masters like Da Vinci, Raphael, and others, a certain amount of artistic skill is not only involved, but required. The same can be said for the bridge builder, the house builder, the car manufacturer, and many other types of builders; in a sense, they are all artisans.

The same can be said for the troubleshooter. While troubleshooting is an art, it is also a skill. The difference between the natural ability of the artist and the skill of the troubleshooter is that the ability to perform correct, accurate troubleshooting can be learned, taught, and mastered—even by those who are not necessarily naturally artistic.

Troubleshooting is defined as the "art" or science of problem solving. Some would simplify this definition by stating that troubleshooting is a simple, systematic method to identify and correct problems. This may be the case; however, experience has shown that nothing is simple and/or systematic unless a certain amount of experience is thrown into the equation.

Another factor needed in the equation is common sense. You may get by (to a degree) in troubleshooting with a lack of expertise, experience, and/or a simplistic, systematic approach—but you can't get by without a great deal of common sense.

Consider the technical student who spends several years in formal classroom training to become an electrician, for example. This student may have little difficulty in understanding Ohm's Law, AC/DC Theory, Inductance/Capacitance, Boolean Algebra, Logic Circuits, and even more

[30]Adapted from Spellman, F. R., *Spellman's Standard Handbook for Wastewater Operators, Volume 2: Intermediate Level*. Lancaster, PA: Technomic Publishing Co., Inc., pp. 53–59, 1999.

complex electrical theory. This same student may ace these subjects and the course itself. But does the ability to score high on theoretical concepts translate to a high degree of proficiency in troubleshooting an electronic /electrical circuit? It depends on experience. As the old adage says, "There simply is no substitute for experience."

If this same brilliant student is exposed to real life (on-the-job) situations where he or she has to troubleshoot complicated systems and solve problems, then over time, he or she will learn. If you add the student's high degree of common sense into this learning experience, then he/she will learn easier and quicker and will retain what he or she has learned.

9.3 TROUBLESHOOTING: WHAT IS IT?

Troubleshooting is the "art" of problem solving. More specifically, troubleshooting provides a mechanism or means to address problems and evaluate possible solutions. It is important to remember that no machine, electrical circuit, or water/wastewater treatment plant unit process runs at maximum efficiency at all times. In fact, in both water and wastewater treatment, unit process problems are not uncommon. Some of the problems are due to poor design or unusual raw water or influent characteristics. However, one thing is absolutely certain: The majority of treatment plant problems are avoidable with better operation, management, maintenance, and process control.

To achieve optimum operation and control, the operator must be able to rapidly identify problems. More importantly, once identified, the problems must be corrected. Maintaining optimum performance of "any" system, including centrifugal pumps, requires the ability to troubleshoot—the ability to troubleshoot correctly.

IMPORTANT

Important Note: *It is important to recognize that, due to the complexity of treatment systems, unit processes, and machinery and equipment such as centrifugal pumping systems and the number of variables involved, troubleshooting may not always identify one single "right" answer. The process requires experience, time, common sense, and usually a good deal of effort to solve performance problems. Probably the most important factor in troubleshooting,* *however, is system knowledge. Simply put, you can't determine what the solution to a problem is with most complex processes, systems, or equipment unless you understand the operation of the process or system or equipment. To remedy difficulties experienced with any treatment unit process, system, or equipment, the operator must know the characteristics of the system or wastestream, the design shortcomings of the process, system, or equipment, and the indicators of operational problems.*

9.4 GOALS OF TROUBLESHOOTING

There are many different reasons for troubleshooting a centrifugal pump. The reasons vary from developing procedures to **prevent** future problems, to improving overall plant performance, and to reducing operation and maintenance costs. However, probably the most common and most important reason for identifying and correcting a problem is when the problem causes the plant to operate at below optimum level that may lead to a permit violation. Centrifugal pumps are part of the treatment process; thus, when they are not operating as designed, they can and will affect process operations. As an aside, it is important to remember that centrifugal pumps may run for years without problems, but some may exhibit problems as soon as startup.

9.5 THE TROUBLESHOOTING PROCESS

IMPORTANT

Note: *There is no one "perfect" method for troubleshooting treatment process problems; each situation can (and probably will) vary. However, certain steps or elements in the process of troubleshooting should be followed each time—these are listed in this section. Keep in mind that these common elements apply to almost all troubleshooting applications, and not just to equipment such as centrifugal pumps.*

The common elements in successful troubleshooting approaches include the following:

- **Observe/Gather Information**—To begin the correct troubleshooting process or approach, there is a need to understand "what is or is not happening" with the process. What doesn't work, or what doesn't work correctly? These questions must be asked first, before any corrective action is attempted. Follow the old adage: "Don't leap before you look."
- **Identify Additional Data Needed**—Many wastewater treatment systems and/or unit processes are quite complex. Simply because one element of the process has been identified as non-functional doesn't necessarily mean that it is the causal factor of the overall malfunction. Many times, failure of one component is caused by failure of some other device or system downstream or upstream of the component not functioning as designed. Collect "all" the information, not just patchwork details.
- **Evaluate Available Information**—Once all the data and information are collected, the next step is to evaluate what you know.
- **Identify Potential Problems, Causes, and Corrective Actions**—With the proper information in hand, the next step is to look at each potential causal factor and then, through process of elimination, narrow down the causes and corrective actions.
- **Prioritize Problems, Causes, and Corrective Actions**—With the potential causes and corrective actions narrowed down to a "small" list, then prioritize them.
- **Select the Actions to Be Taken**—A short list of actions to take should be made to determine if any of them are the corrective action(s) required.
- **Implement**—Starting with the most likely corrective action, implement it first, and then proceed down the list until the problem is remedied. (Note: If the problem is not remedied from the corrective actions listed, then start the process over from the beginning.)
- **Observe the Results**—Once the selected remedial action is effected, observing the impact of the action is important, obviously.
- **Documenting (Recordkeeping) for Future Use**—One thing is certain: If a system or unit process fails once, it is quite likely that it will

fail again. If this is the case, then why would anyone want to spend an inordinate amount of time and effort finally determining the causal factor without taking the time to document it? Local knowledge is important in maintaining the smooth, uninterrupted operation of a water/wastewater treatment plant; local troubleshooting knowledge is more than important, it is critical.

A simplified basic summary of the elements listed above is shown in Figure 9.1. Figure 9.2 shows a more detailed step-by-step troubleshooting process. It is a model or paradigm that has one huge advantage over many others: it has been tested.

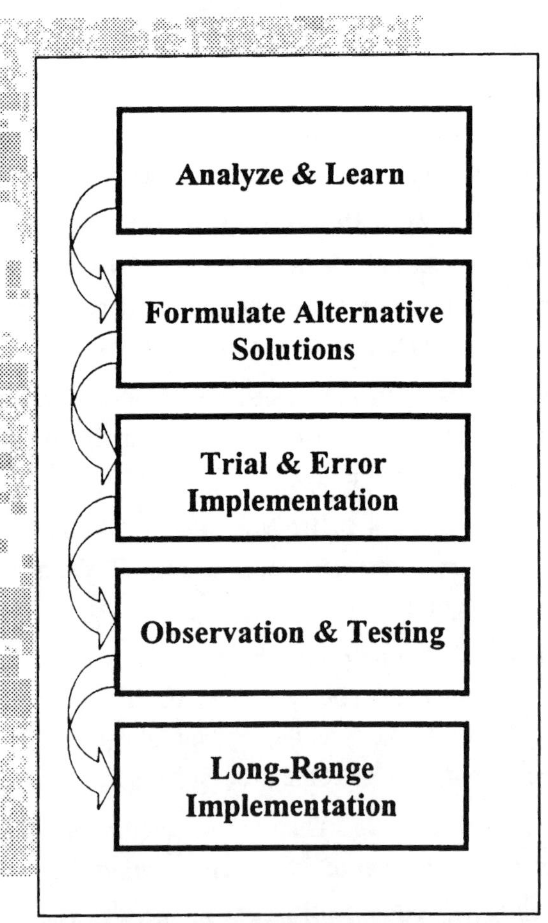

Figure 9.1
Simplified troubleshooting steps.

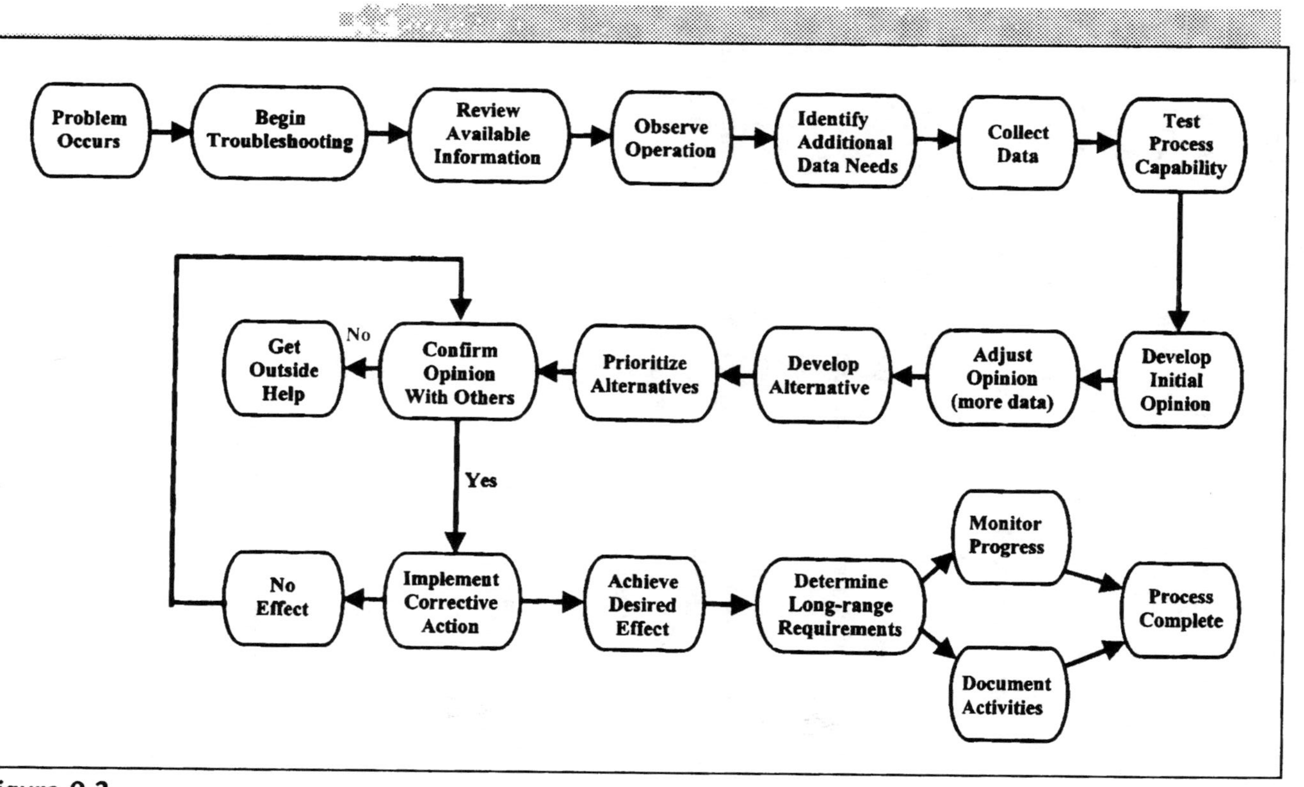

Figure 9.2
Troubleshooting sequence.

9.6 TROUBLESHOOTING THE CENTRIFUGAL PUMP

We stated earlier that anyone attempting to troubleshoot a system, a process, or a piece of equipment without first knowing the basic operation of the system, process, and/or equipment is performing a foolhardy exercise at best. Accordingly, the maintenance operator (the troubleshooter) must know the characteristics, the design shortcomings, and the indicators of operational problems for the centrifugal pump he or she is attempting to troubleshoot.

The following sections provide 13 typical problems associated with centrifugal pumps. Along with the problem or symptom, possible causes and the required action or remedy are also provided.

Important Note: *Again, it is important to remember that the manufacturer's technical manual for the centrifugal pump is the resource that should be used in troubleshooting. Moreover, only experienced maintenance operators should perform the actual troubleshooting process.*

9.6.1 PUMP FAILS TO PRIME OR LOSES ITS PRIME

Possible Cause:

1. There are air leaks in the suction line.
2. Suction strainer or inlet is clogged.
3. Suction lift is too high.
4. Priming unit is defective.
5. Defective packing or seal is causing air leakage.
6. There is excessive air or vapor in liquid.
7. There is an air pocket in the suction line
8. Air leaks into pump through gaskets.
9. Inlet or suction line is improperly located.

10. Lantern ring is improperly located.
11. Water seal pipe is plugged

Action or Remedy:

1. Inspect, clean, and tighten all suction connections.
2. Remove dirt, leaves, or other material from strainer or inlet.
3. Reevaluate pump requirements, and correct suction conditions accordingly.
4. Inspect, clean, and repair priming unit; replace any defective parts.
5. Inspect packing or seal for proper operation; listen and feel for evidence of air leaking into pump through the stuffing box; check shaft and its sleeve for scoring.
6. Visually observe liquid being pumped for evidence of excessive air or vapor concentrations.
7. Open air bleed-off valves in suction piping or on pump; listen for excessive noise in suction lines.
8. Visually inspect the pump's gasket and O-ring locations for signs of cracks or deterioration; feel around these locations for signs of air leakage.
9. Reevaluate suction line configurations, and determine if they need to be relocated or modified.
10. Consult manufacturer's literature, and inspect stuffing box to ensure proper location of packing and lantern ring.
11. Inspect external water seal unit; remove piping at pump end to ensure water is flowing, and inspect connection at the stuffing box.

9.6.2 PUMP DOES NOT DISCHARGE

Possible Cause:

1. Pump is not properly primed.
2. Total head is too high.
3. Drive unit is not operating at rated speed.

4. Impeller or discharge line is clogged.
5. Suction or discharge check or line valves are closed.
6. Wrong direction of rotation.
7. Pump is vapor bound.
8. Suction line location problems.
9. Suction lift too high.
10. Air leaks in pump or suction piping.

Action or Remedy:

1. Reprime the pump; refer to priming troubleshooting and remedies.
2. Reevaluate head calculations; measure elevation differences between pump and liquid source, and pump and discharge points; check pipe friction losses; are valves wide open?
3. Check voltage of electric motor; check steam pressure of steam turbine; check engine rpms. Refer to applicable maintenance manuals for possible trouble and corrective action.
4. Backflush pump to try to clear obstruction; remove pump inspection plate (if so equipped) and clean impeller; dismantle pump and/or piping and remove obstruction.
5. Check all suction and discharge line valves associated with the pump to ensure proper positioning; inspect and operate check valves to ensure proper operation.
6. Observe pump rotation, and compare with indicator arrow; check wiring against diagram on motor name plate and in controller; consult manufacturer's literature for troubleshooting procedures; acquire the services of a qualified electrician to perform needed repairs.
7. Provide additional pressure on liquid being pumped by elevating liquid level; bleed excess air from pump and suction piping.
8. Reevaluate suction line configurations, and determine if they need to be relocated or modified.
9. Reevaluate pump requirements, and correct suction conditions accordingly.
10. Inspect, clean, and tighten all suction connections. Inspect packing or seal for proper operation; listen and feel for evidence of air leaking into pump through the stuffing box; check shaft and its sleeve for

scoring. Visually inspect the pump's gasket and O-ring locations for signs of cracks or deterioration; feel around these locations for signs of air leakage.

9.6.3 PUMP DOES NOT DELIVER RATED CAPACITY

Possible Cause:

1. Pump is not properly primed.
2. Suction lift is too high.
3. Excessive air or vapor in liquid.
4. Air leaking through stuffing box, suction line, and/or pump casing.
5. Drive unit is not operating at rated speed.
6. Impeller is clogged.
7. Wear rings are worn.
8. Impeller is damaged or eroded.
9. Pump is vapor bound.
10. Discharge pressure required by the system is greater than that for which the pump was designed.
11. Insufficient suction head (cavitation occurring).
12. Foot valve too small or clogged.
13. Vortexing.
14. Air pocket in suction line.
15. Water seal pipe plugged.

Action or Remedy:

1. Reprime the pump; refer to priming troubleshooting and remedies.
2. Reevaluate pump requirements, and correct suction conditions accordingly.
3. Inspect packing or seal for proper operation; listen and feel for evidence of air leaking into pump through the stuffing box; check shaft and its sleeve for scoring.
4. Visually observe liquid being pumped for evidence of excessive air or

vapor concentrations. Inspect, clean, and tighten all suction connections. Visually inspect the pump's gasket and O-ring locations for signs of cracks or deterioration; feel around these locations for signs of air leakage.

5. Check voltage of electric motor; check steam pressure of steam turbine; check engine rpms. Refer to applicable maintenance manuals for possible trouble and corrective action.
6. Backflush pump to try to clear obstruction; remove pump inspection plate (if so equipped) and clean impeller; dismantle pump and/or piping and remove obstruction.
7. Inspect pump wear rings for excessive wear; measure clearances, and, if out of the acceptable range, replace the wear rings.
8. Remove pump inspection plate (if so equipped) or disassemble pump, and inspect impeller; if damaged or eroded, replace impeller.
9. Provide additional pressure on liquid being pumped by elevating liquid level; bleed excess air from pump and suction piping.
10. Reevaluate pumping system requirements, and make the necessary modifications or changes.
11. Increase the liquid level in the pumped tank; reevaluate the design of the suction portion of the pumping system.
12. Inspect the foot valve; free area through all parts of the valve should be no less than 1.5 times the area of the suction pipe; if a strainer is used, area should be three or four times the area of the suction pipe.
13. Increase the liquid level in the pumped tank; reevaluate the design of the suction portion of the pumping system.
14. Open air bleed-off valves in suction piping or on pump; listen for excessive noise in suction lines.
15. Inspect external water seal unit; remove piping at pump end to ensure water is flowing, and inspect connection at the stuffing box.

9.6.4 PUMP DOES NOT DELIVER SUFFICIENT PRESSURE

Possible Cause:

1. Excessive air or vapor is in liquid.

2. Drive unit is not operating at rated speed.
3. Direction of rotation is wrong.
4. Total head is too high.
5. Wear rings are worn.
6. Impeller is damaged or eroded.
7. Air leaks through pump or suction lines.
8. Suction or discharge valves are partially opened.
9. Suction or discharge lines are partially clogged.
10. Impeller size is not correct.

Action or Remedy:

1. Visually observe liquid being pumped for evidence of excessive air or vapor concentrations.
2. Check voltage of electric motor; check steam pressure of steam turbine; check engine rpms. Refer to applicable maintenance manuals for possible troubles and corrective action.
3. Observe pump rotation, and compare with indicator arrow; check wiring against diagram on motor name plate and in controller; consult manufacturer's literature for troubleshooting procedures; acquire the services of a qualified electrician to perform needed repairs.
4. Reevaluate head calculations; measure elevation differences between pump and liquid source, and pump and discharge points; check pipe friction losses; determine whether valves are wide open.
5. Inspect pump wear rings for excessive wear; measure clearances, and, if out of the acceptable range, replace the wear rings.
6. Remove pump inspection plate (if so equipped), or disassemble pump and inspect impeller; if damaged or eroded, replace impeller.
7. Inspect, clean, and tighten all suction connections. Inspect packing or seal for proper operation; listen and feel for evidence of air leaking into pump through the stuffing box; check shaft and its sleeve for scoring. Visually inspect the pump's gasket and O-ring locations for signs of cracks or deterioration; feel around these locations for signs of air leakage.
8. Check all suction and discharge line valves associated with the pump

to ensure proper positioning; inspect and operate check valves to ensure proper operation.

9. Remove dirt, leaves, or other material from strainer or inlet, and/or backflush the pump if possible; if not, dismantle the pump, and remove any obstructions.
10. Check actual diameter of impeller; consult with pump manufacturer to see if impeller needs to be replaced.

9.6.5 PUMP STARTS AND STOPS PUMPING

Possible Cause:

1. Air leaks through pump or suction lines.
2. There is an air pocket in suction line.
3. Excessive air or vapor is in liquid.
4. Water seal pipe or line is plugged.
5. Suction lift is too high.
6. Packing or mechanical seal is defective.
7. Pump is vapor bound.
8. Pump was not properly primed.
9. Casing is distorted by excessive strains from suction or discharge piping.
10. Shaft is bent due to thermal distortion, damage during overhaul, or improper assembly of rotating elements.
11. There is mechanical failure of critical pump parts.

Action or Remedy:

1. Inspect, clean, and tighten all suction connections. Inspect packing or seal for proper operation; listen and feel for evidence of air leaking into pump through the stuffing box; check shaft and its sleeve for scoring. Visually inspect the pump's gasket and O-ring locations for signs of cracks or deterioration; feel around these locations for signs of air leakage.
2. Open air bleed-off valves in suction piping or on pump; listen for excessive noise in suction lines.

3. Visually observe liquid being pumped for evidence of excessive air or vapor concentrations.
4. Inspect external water seal unit; remove piping at pump end to ensure water is flowing, and inspect connection at the stuffing box.
5. Reevaluate pump requirements, and correct suction conditions accordingly.
6. Inspect packing or seal for proper operation; listen and feel for evidence of air leaking into pump through the stuffing box; check shaft and its sleeve for scoring.
7. Provide additional pressure on liquid being pumped by elevating liquid level; bleed excess air from pump and suction piping.
8. Reprime the pump; refer to priming troubleshooting and remedies.
9. Inspect interior of pump for friction and wear between impeller and casing; replace damaged parts and eliminate piping strains.
10. Check shaft deflection by turning between lathe centers; shaft warpage should not exceed 0.002" on any pump.
11. Inspect bearings and impeller for damage; any irregularity on the parts will cause a drag on the shaft and improper operation.

9.6.6 PUMP OVERLOADS DRIVER OR CONSUMES EXCESSIVE POWER

Possible Cause:

1. Motor speed is too high.
2. Direction of rotation is wrong.
3. Total head is too high.
4. Total head is too low.
5. Impeller is clogged.
6. Impeller size is not correct.
7. Motor shaft is bent.
8. Drive unit and pump are misaligned.
9. Wear rings are worn.
10. Packing is improperly installed.
11. Rotating and stationary parts are rubbing.

Action or Remedy:

1. Internal electric motor wiring is incorrect; replace motor; refer to applicable drive unit maintenance manuals for possible trouble and corrective action.
2. Observe pump rotation, and compare with indicator arrow; check wiring against diagram on motor name plate and in controller; consult manufacturer's literature for troubleshooting procedures; acquire the services of a qualified electrician to perform needed repairs.
3. Reevaluate head calculations; measure elevation differences between pump and liquid source, and pump and discharge points; check pipe friction losses; determine whether valves are wide open.
4. Reevaluate head conditions; correct as required.
5. Backflush pump to try to clear obstruction; remove pump inspection plate (if so equipped), and clean impeller; dismantle pump and/or piping, and remove obstruction.
6. Check actual diameter of impeller; consult with pump manufacturer to see if impeller needs to be replaced.
7. Inspect motor shaft; replace shaft if it's bent.
8. Inspect shafts and couplings for angular and parallel misalignment; if out of alignment, correct the misalignment.
9. Inspect pump wear rings for excessive wear; measure clearances, and, if out of the acceptable range, replace the wear rings.
10. Remove packing; inspect and reinstall correctly; replace packing if necessary.
11. Rotate pump by hand listening and feeling for rubbing parts; if impeller is binding, realign or relieve strain on casing; adjust impeller clearance; replace worn or damaged parts.

9.6.7 PUMP IS NOISY OR HAS EXTENSIVE VIBRATION

Possible Cause:

1. There is a magnetic hum.

2. Motor bearings are worn.
3. Impeller is clogged.
4. Impeller is binding.
5. Motor shaft is bent or worn.
6. Drive and pump are misaligned.
7. Foundation is not rigid.
8. Impeller is damaged.
9. Pump is not properly leveled.
10. Piping is not supported.
11. Pump is cavitating.
12. Pump or suction pipe is not completely filled with liquid.
13. Foot valve is too small or partially clogged.
14. Pump shaft is bent.
15. There is excessive lubrication in motor and pump ball bearings.
16. There is a lack of lubrication in motor and pump ball bearings.
17. There is internal misalignment due to worn bearings.

Action or Remedy:

1. Consult motor manufacturer's technical manual.
2. Replace bearings.
3. Backflush pump to try to clear obstruction; remove pump inspection plate (if so equipped), and clean impeller; dismantle pump and/or piping, and remove obstruction.
4. Rotate pump by hand listening and feeling for rubbing parts; if impeller is binding, realign or relieve strain on casing; adjust impeller clearance; replace worn or damaged parts.
5. Inspect motor shaft; replace shaft if it's bent.
6. Inspect shafts and couplings for angular and parallel misalignment; if out of alignment, correct the misalignment.
7. Inspect foundation; strengthen or change method of mounting pump unit.
8. Remove pump inspection plate (if so equipped), or disassemble pump and inspect impeller; if damaged or eroded, replace impeller.

9. Check levelness of pump; make necessary changes to relevel the pump; recheck shaft alignments.
10. Provide support for suction and discharge piping.
11. Inspect pump impeller casing for signs of cavitation; if present, reevaluate pump application and suction and discharge piping; consult with pump manufacturer.
12. Open air purges on suction line and pump; reevaluate suction piping and pump location; modify system as needed.
13. Inspect the foot valve; free area through all parts of the valve should be no less than 1.5 times the area of the suction pipe; if a strainer is used, area should be three or four times the area of the suction pipe.
14. Inspect bearings and impeller for damage; any irregularity on the parts will cause a drag on the shaft and improper operation.
15. Inspect bearings; clean with solvent, and relubricate properly; replace if damaged.
16. Inspect bearings; clean with solvent, and relubricate properly; replace if damaged.
17. Replace bearings.

9.6.8 PACKING HAS A SHORT LIFE

Possible Cause:

1. Water seal pipe plugged or no water being provided
2. Seal cage improperly located
3. Packing improperly installed
4. Incorrect packing for operating conditions
5. Packing gland too tight
6. Dirt or grit in sealing liquid
7. Shaft misalignment
8. Shaft bent
9. Bearings worn
10. Shaft sleeve worn or scored

Action or Remedy:

1. Inspect external water seal unit; remove piping at pump end to ensure water is flowing, and inspect connection at the stuffing box.
2. Consult manufacturer's literature, and inspect stuffing box to ensure proper location of packing and lantern ring.
3. Remove packing; inspect and reinstall correctly; replace packing if necessary.
4. Consult with manufacturer's representative or read literature and select the proper packing.
5. Remove gland and packing; repack the pump, and follow the recommended procedures for adjusting the gland.
6. Add filters to the sealing liquid line; use a clean water source for sealing liquid.
7. Inspect shafts and couplings for angular and parallel misalignment; if out of alignment, correct the misalignment.
8. Inspect motor shaft; replace shaft if it's bent.
9. Replace bearings.
10. Remove gland and packing, and inspect shaft sleeve for wear. This generally has to be done with the hands or a packing tool.

9.6.9 STUFFING BOX LEAKS EXCESSIVELY

Possible Causes:

1. Refer to possible causes in Section 9.6.8.

Action or Remedy:

1. Refer to remedies in Section 9.6.8.

9.6.10 MECHANICAL SEAL HAS A SHORT LIFE

Possible Causes:

1. Shaft bent

2. Shaft sleeve worn or scored
3. Seal improperly installed
4. Improper seal for operating conditions
5. Abrasive solids in liquid being pumped
6. Mechanical seal run dry
7. Bearings worn
8. Pump shaft misalignment

Action or Remedy:

1. Inspect motor shaft; replace shaft if it's bent.
2. Remove gland and packing, and inspect shaft sleeve for wear. This generally has to be done with the hands or a packing tool.
3. Remove seal; consult with manufacturer's representative, and read instructions with seal; reinstall seal following proper instructions.
4. Consult with manufacturer's representative, and choose proper seal for application.
5. Provide a separate clean seal water source.
6. Operate pump only while seal water is flowing.
7. Replace bearings.
8. Inspect shaft for damage due to misalignment; replace if damaged; correct problem causing misalignment (worn bearings, out of balance impeller, pipe strain, etc.).

9.6.11 MECHANICAL SEAL LEAKS EXCESSIVELY

Possible Cause:

1. Leakage occurs under shaft sleeve due to gasket or O-ring failure.
2. Refer to possible causes in Section 9.6.10.

Action or Remedy:

1. Determine if leakage is actually between shaft sleeve and shaft; replace gasket or O-ring.
2. Inspect motor shaft; replace shaft if it's bent.

3. Remove gland and packing, and inspect shaft sleeve for wear. This generally has to be done with the hands or a packing tool.
4. Remove seal; consult with manufacturer's representative, and read instructions with seal; reinstall seal following proper instructions.
5. Consult with manufacturer's representative, and choose proper seal for application.
6. Provide a separate clean seal water source.
7. Operate pump only while seal water is flowing.
8. Replace bearings.
9. Inspect shaft for damage due to misalignment; replace if damaged; correct problem causing misalignment (worn bearings, out of balance impeller, pipe strain, etc.).

9.6.12 BEARINGS HAVE A SHORT LIFE

Possible Cause:

1. Bent or damaged shaft
2. Excessive thrust caused by a mechanical failure inside the pump
3. Excessive grease in bearings
4. Lack of lubrication in bearings
5. Rusting of bearings due to water getting into bearings past slinger ring
6. Out of balance shaft or impeller

Action or Remedy:

1. Check shaft deflection by turning between lathe centers; shaft warpage should not exceed 0.002" on any pump.
2. Dismantle pump, and determine the extent of the damage; if capable and the right tools are available, repair internal damage.
3. Replace bearing; relubricate following accepted procedures.
4. Replace bearing; lubricate following accepted procedure.
5. Replace and relubricate damaged bearing; replace or repair slinger ring.
6. Replace shaft or impeller.

9.6.13 PUMP OVERHEATS AND/OR SEIZES

Possible Cause:

1. Pump not primed and allowed to run dry
2. Vapor or air pockets inside of pump
3. Operation at too low capacity
4. Internal misalignment due to improper repairs
5. Rubbing on rotating and stationary parts
6. Worn bearings
7. Lack of lubrication

Action or Remedy:

1. Check bearing temperature, and try to turn pump by hand; if pump turns freely, try repriming pump again; if it does not turn freely, inspect pump for impeller damage, bent shaft, and/or seized bearings.
2. Shut the pump down; check the bearing housing and volute case for excessive temperature conditions; turn the pump by hand; if pump turns freely, start the pump, and open the air release plugs to allow entrapped air out; if pump does not turn freely, inspect pump impeller, shaft, volute, and bearings for damage.
3. Reevaluate pump system operating conditions; consult with pump supplier for corrective actions.
4. Shut the pump down; turn pump by hand; if pump does not turn freely or drags, inspect pump impeller, shaft, and wear rings for damage; have competent repairman service the pump.
5. See remedy in Section 9.6.13, number 4.
6. Replace bearings.
7. Shut the pump down; check bearing temperature; turn the pump by hand; if pump turns freely, relubricate the bearings according to lubrication instructions, and follow a prescribed lubrication schedule in the future; if pump has seized, replace and lubricate bearings; turn pump by hand once again to ensure it is ready to run.

REFERENCES

Spellman, F. R., *Spellman's Standard Handbook for Wastewater Operators, Volume 2: Intermediate Level.* Lancaster, PA: Technomic Publishing Co., Inc., 1999.

Worldwide Pump Market Expected to Reach $22 Billion by 2003. *U.S. Water News,* 17, 2, February 2000.

Self-Test

9.1 In a paragraph or less, explain the primary elements that contribute to an operators's ability to properly troubleshoot centrifugal pump problems.

Centrifugal Pump: Modifications

The centrifugal pump (and its modifications) is one of the most diversified hydraulic machines used in water or wastewater treatment.

TOPICS

Submersible Pumps
Recessed Impeller or Vortex Pumps
Turbine Pumps

10.1 INTRODUCTION

Along with its reliability and other advantages, the centrifugal pump can also be used in a wide range of different applications.

The centrifugal pump's wide diversity in application is the result of its adaptability through various modifications. These modifications include variations in impeller design, including the use of semi-open, open, or closed impellers; vertical or horizontal shaft configuration; and different priming mechanisms, such as conventional flooded-suction, self-priming, and vacuum-priming. Each of these variations enables pumps to meet certain design criteria, including head, capacity, and efficiency requirements, while ensuring maintenance accessibility, eliminating pump clogging, and accommodating piping needs. One example is if there is a need to produce higher discharge heads, the pump may be modified to include several additional impellers. Another example is if the material being pumped contains a large amount of material that could clog the pump, the pump construction may be modified to remove the major portion of the impeller from direct contact with material being pumped.

Numerous modifications of the centrifugal pump are available, however, the scope of this handbook covers only those that have found wide application in the water/wastewater treatment industries.

Key Terms Used in This Chapter

SUBMERSIBLE PUMP	Vertical, close-coupled centrifugal pump that is designed to work with both motor and pump submerged in the water being pumped.
SUBMERSIBLE TURBINE PUMP	A style of vertical turbine pump in which the entire pump assembly and motor are submersed in the water. The motor is commonly mounted below the pump.

Modifications presented in this chapter include:

- submersible pumps
- recessed impeller or vortex pumps
- turbine pumps

10.2 SUBMERSIBLE PUMPS

The *submersible pump* is, as the name suggests, placed directly in deep wells and in wet wells. In some cases, only the pump is submerged, while in other cases, the entire pump-motor assembly is placed in the well or wet well. A simplified diagram of a typical submersible pump is shown in Figure 10.1.

The submersible pump may be either a close-coupled centrifugal pump or an extended shaft centrifugal pump. If the system is a close-coupled pump system, then both motor and pump are submerged in the water/wastewater being pumped. Seals to prevent the water/wastewater from entering the inside of the motor causing shorts and motor burnout must protect the electric motor in a close-coupled pump.

In the extended shaft system, the pump is submerged while the motor is mounted above the pump well or wet well. In this situation, the pump and motor must be connected by one of the extended shaft assemblies discussed earlier.

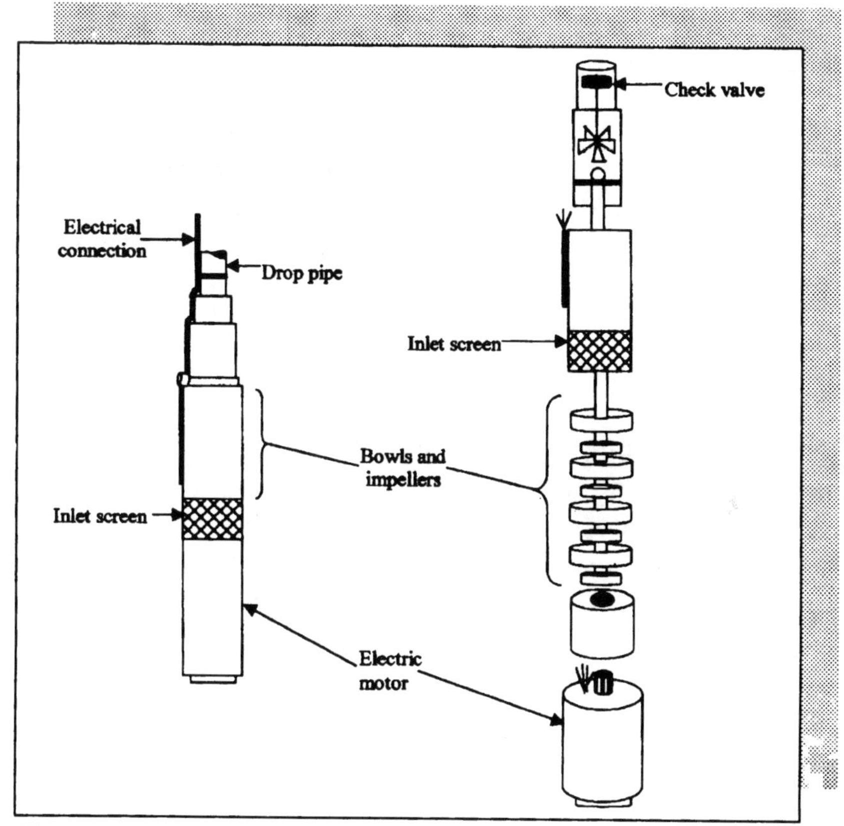

Figure 10.1
Submersible pump.

10.2.1 APPLICATIONS

The submersible pump has wide applications in the water and/or wastewater treatment industry. It generally can be substituted in any of the applications of other types of centrifugal pumps. However, it has found its widest application in collector system pump stations.

10.2.2 ADVANTAGES

In addition to the advantages discussed earlier for a conventional centrifugal pump, the submersible pump has additional advantages:

- Because it is located below the surface of the liquid, there is less chance that the pump will lose its prime, develop air leaks on the suction side of the pump, or require initial priming.
- Because the pump or the entire assembly is located in the well or wet well, there is less cost associated with the construction and operation of this system. There is no need to construct a dry well or a large structure to hold the pumping equipment and necessary controls.

10.2.3 DISADVANTAGES

The major disadvantage associated with the submersible pump is the lack of access to the pump or pump and motor. The performance of any maintenance requires either drainage of the well (not likely) or wet well or extensive lift equipment to remove the equipment from the well (more likely)/wet well or both. This may be a major factor in determining if a pump receives the attention it requires. Also, in most cases, all major maintenance on close-coupled submersible pumps must be performed by outside contractors due to the need to reseal the motor to prevent leakage.

10.3 RECESSED IMPELLER OR VORTEX PUMPS

The *recessed impeller* or *vortex pump* uses an impeller that is either partially or wholly recessed into the rear of the casing (see Figure 10.2). The spinning action of the impeller creates a vortex or whirlpool. This whirlpool increases the velocity of the material being pumped. As in other centrifugal pumps, this increased velocity is then converted to increased pressure or head.

10.3.1 APPLICATIONS

The recessed impeller or vortex pump is used widely in applications where the liquid being pumped contains large amounts of solids or debris (e.g., wastewater sludge) that could clog or damage the pump's impeller.

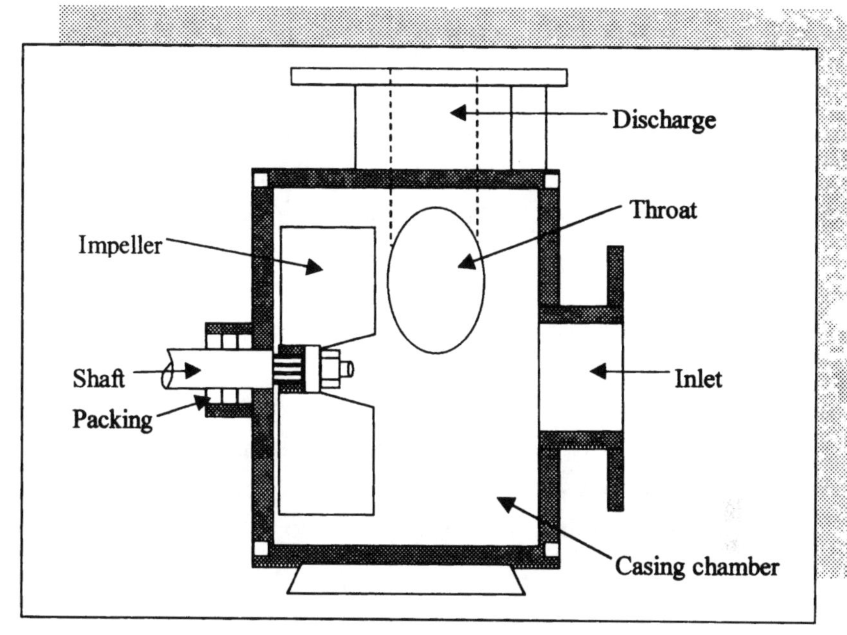

Figure 10.2
Schematic of a recessed impeller or vortex pump.

It has found increasing use as a sludge pump in facilities that withdraw sludge continuously from their primary clarifiers.

10.3.2 ADVANTAGES

The major advantage of this modification is the increased ability to handle materials that would normally clog or damage the pump impeller. Because the majority of the flow does not come in direct contact with the impeller, there is much less potential for problems.

10.3.3 DISADVANTAGES

Because there is less direct contact between the liquid and the impeller, the energy transfer is less efficient. This results in somewhat higher power costs and limits the pump's application in low to moderate capacities.

Objects that might have clogged a conventional-type centrifugal pump are now able to pass through the pump. Although this is very beneficial in reducing pump maintenance requirements, it can, in some situations, allow material to pass into a less accessible location before becoming an obstruction. To be effective, the piping and valving must be designed to pass objects of a size equal to that which the pump will discharge.

10.4 TURBINE PUMPS

The *turbine pump* consists of a motor, drive shaft, a discharge pipe of varying lengths, and one or more impeller-bowl assemblies. It is normally a vertical assembly in which the water enters at the bottom, passes axially through the impeller-bowl assembly where the energy transfer occurs, then moves upward through additional impeller-bowl assemblies to the discharge pipe. The length of this discharge pipe will vary with the distance from the wet well to the desired point of discharge (see Figure 10.3).

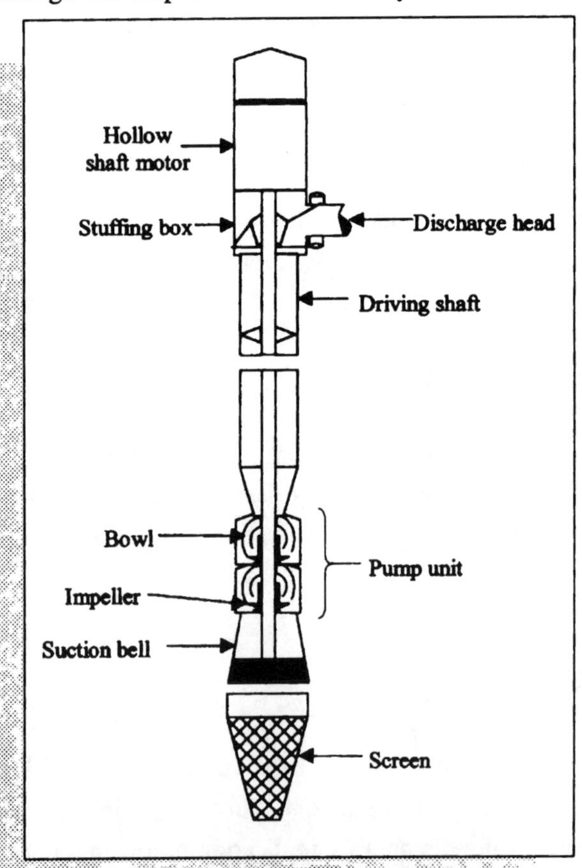

Figure 10.3
Vertical turbine pump.

There are two basic types of turbine pumps:

- line shaft turbines
- can turbine (for dry well installations)

10.4.1 APPLICATION

Due to the construction of the turbine pump, the major applications have traditionally been for pumping relatively clean water. The line shaft turbine pump has been used extensively for drinking water pumping, especially where water is withdrawn from deep wells. The main wastewater treatment application has been pumping plant effluent back into the plant for use as service water.

10.4.2 ADVANTAGES

The turbine pump has a major advantage in the amount of head it is capable of producing. By installing impeller-bowl assemblies, the pressure or head the pump is capable of producing may be increased dramatically.

Important Point: *The amount of pressure in a lineshaft turbine pump can be increased by adding impeller-bowl assemblies.*

10.4.3 DISADVANTAGES

The presence of large amounts of solids within the liquid being pumped can seriously increase the amount of maintenance the pump requires. As a result, in wastewater treatment, the unit has not found widespread use in any other situation other than service water pumping.

Self-Test

10.1 The amount of pressure in a lineshaft turbine pump can be increased by adding ___________.

10.2 Describe the recessed impeller-type centrifugal pump.

10.3 What are the advantages of the recessed impeller?

10.4 Describe the submersible-type centrifugal pump.

10.5 What are the advantages of the submersible pump?

10.6 What is the main use of the lineshaft turbine in a wastewater treatment plant?

Positive-Displacement Pumps

Early water systems used reciprocating positive-displacement pumps powered by steam engines to obtain the pressure needed to supply water to customers. These pumps have essentially all been replaced with centrifugal pumps, which are much more efficient. The only types of positive-displacement pumps used in current water [and wastewater] systems are some types of portable pumps used to dewater excavations, pumps used to convey sludge/biosolids, as well as chemical feed pumps.[31]

11

Positive-Displacement Pumps

TOPICS

Reciprocating Pumps
Rotary Pumps
Special-Purpose Pumps

11.1 INTRODUCTION

The clearest differentiation between centrifugal (or kinetic) pumps and positive-displacement pumps is made or based on the method by which pumping energy is transmitted to the liquid. As pointed out earlier, kinetic (centrifugal) pumps rely on a transformation of kinetic energy to static pressure. Positive-displacement pumps, on the other hand, discharge a given volume for each stroke or revolution (i.e., energy is added intermittently to the fluid flow).

The two most common forms of positive-displacement pumps are reciprocating action pumps (which use pistons, plungers, diaphragms, or bellows) and rotary action pumps (using vanes, screws, lobes, or pro-

[31]AWWA, *Water Transmission and Distribution,* 2nd ed. Denver: American Water Works Association, p. 369, 1996.

Key Terms Used in This Chapter

DIAPHRAGM PUMP	Positive-displacement pump of the plunger type that employs a flexible diaphragm as the pumping mechanism.
PERISTALTIC PUMP	As the fluid passes through flexible tubing, an external moving element progressively flattens the tubing, pushing the liquid forward.
POSITIVE-DISPLACEMENT PUMP	The fluid is forced to move because it is displaced by the movement of a piston, vane, screw, or roller. Positive-displacement pumps act to force water into a system regardless of the resistance that may oppose the transfer.
PROGRESSIVE-CAVITY PUMP	A positive-displacement pump in which a rotary motion opens a cavity that moves forward along the pump length.
ROTARY GEAR PUMP	Inter-meshing gears trap the liquid and move it from suction to discharge.

gressing cavities). Regardless of form used, all positive-displacement pumps act to force liquid into a system regardless of the resistance that may oppose the transfer. The discharge pressure generated by a positive-displacement pump is, in theory, infinite. If the pump is deadheaded, the pressure generated will increase until either a pump part fails or the driver stalls from lack of power.[32]

***Important Point:** Because positive-displacement pumps **cannot** be operated against a closed discharge valve (i.e., something must be displaced with each stroke of the pump), closing the discharge valve can cause rupturing of the discharge pipe, the pump head, the valve, or some other component.*

The three basic types of positive-displacement pumps discussed in this chapter are

- reciprocating pumps

[32]Wahren, U., *Practical Introduction to Pumping Technology.* Houston: Gulf Publishing Company, p. 30, 1997.

- rotary pumps
- special-purpose pumps (peristaltic or tubing pumps)

11.2 RECIPROCATING PUMPS

The *reciprocating pump* (or piston pump) is one type of positive-displacement pump. This pump works just like the piston in an automobile engine—on the intake stroke, the intake valve opens, filling the cylinder with liquid. As the piston reverses direction, the intake valve is pushed closed, and the discharge valve is pushed open; the liquid is pushed into the discharge pipe. With the next reversal of the piston, the discharge valve is pulled closed, and the intake valve is pulled open; then, the cycle repeats.

A piston pump is usually equipped with an electric motor and a gear and cam system that drives a plunger connected to the piston. Just like an automobile engine piston, the piston must have packing rings to prevent leakage and must be lubricated to reduce friction. Because the piston is in contact with the liquid being pumped, only good grade lubricants can be used when pumping materials that will be added to drinking water. The valves must be replaced periodically as well.

Four major types of reciprocating pumps exist:

- steam pumps
- power pumps
- diaphragm pumps
- metering pumps

In this handbook, only diaphragm and metering pumps are discussed, because of their application in water/wastewater operations.

11.2.1 DIAPHRAGM PUMPS

A *diaphragm pump* is composed of a chamber used to pump the fluid; a diaphragm that is operated by either electric or mechanical means; and two valve assemblies—a suction and a discharge valve assembly (see Figure 11.1).

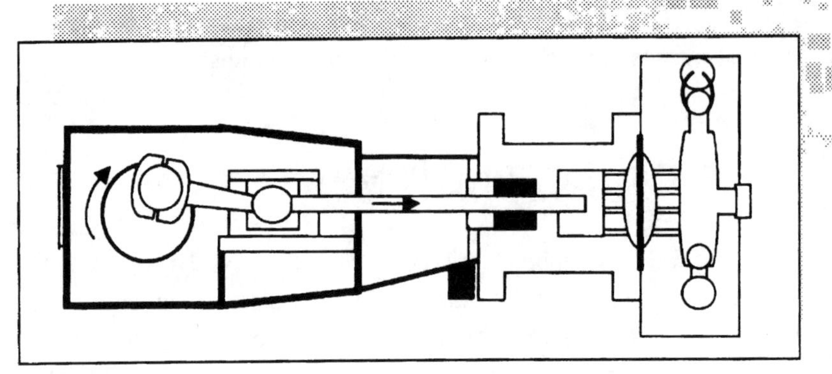

Figure 11.1
Diaphragm pump.

A diaphragm pump is a variation of the piston pump in which the plunger is isolated from the liquid being pumped by a rubber or synthetic diaphragm. As the diaphragm is moved back and forth by the plunger, liquid is pulled into and pushed out of the pump. This arrangement provides better protection against leakage of the liquid being pumped and allows the use of lubricants that otherwise would not be permitted. Care must be taken to ensure that diaphragms are replaced before they rupture. Diaphragm pumps are appropriate for discharge pressures up to about 125 psi, but do not work well if they must lift liquids more than about four feet.

Diaphragm pumps are frequently used for chemical feed pumps. They are well suited for this purpose because the liquid (corrosive liquids, slurries, abrasive liquids, food, or chemicals) only touches the diaphragm, the suction, and the discharge. By adjusting the frequency of the plunger motion and the length of the stroke, extremely accurate flow rates can be metered. The pump may be driven hydraulically by an electric motor or by an electronic driver in which the plunger is operated by a solenoid. Electronically driven metering pumps are extremely reliable (few moving parts) and inexpensive.

11.2.2 METERING PUMPS

To gain an understanding of what a *metering pump* is, what it does, and what it is capable of doing, we must begin by contemplating the first metering pump.

Consider these important points:

- sized to fit the system conditions
- positive-displacement ability for repeatability, reproducibility, and linearity
- flow rate adjustable for varying conditions
- adjustable for varying pressures
- pulsing or reciprocating action
- check valves
- leak proof
- built-in safety features
- relatively simple in construction compared to other parts of the system
- rugged in construction, requiring little maintenance.
- can meter two different liquids at the same time and maintain synchronization
- all parts of the system depend on it
- electrically actuated (This is not necessary for a metering pump, the vast majority of metering pumps are capable of other types of actuation.)[33]

Fortunately, people weren't required to invent the device described above; instead, they were born with it beating inside their chests—it is the human heart, of course. This very first metering pump, in regards to its simplicity/complexity and design and efficiency, is without comparison. But we try, as humans, to duplicate what we know works and works so well. In this instance, we know that the human heart is the absolute embodiment of what a metering pump should be. So, in our quest to design and build the most efficient and practical metering pump we possibly can, we fashion a likeness of, a close resemblance to, a very close correlation to the operation of the ultimate model, the human heart. But, we have not been able to build a machine that matches the human heart's metering function.

[33]From McCabe, R. E., Lanckton, P. G. and Dwyer, W. V., *Metering Pump Handbook.* New York: Industrial Press, Inc., pp. 4–5, 1984.

Figure 11.2
(Top) Shows a meter pump system used to meter sodium hypochlorite in a wastewater treatment plant. (Bottom) Close-up of meter pump assembly.

Let's leave our brief review of anatomy and move on to the less dramatic (but just as relevant) world of metering pumps. More specifically, let's move our focus to metering pumps that are commonly used in water and wastewater treatment operations.

Metering pumps have found a wide range of uses in water/wastewater treatment operations (see Figure 11.2). In fact, their use is without comparison in other industries. Simply, the largest area of application of metering pumps is in water and/or wastewater treatment.

Continuous accurate treatment of water and/or wastewater is required in all phases of its use, reuse, or disposal. Most raw water supplies and wastewater treatment operations are treated with chlorine to control bacteria growth. Some waterworks meter hydrofluosilicic acid to fluoridate the water for improving growth of teeth in children. Metering pumps are used to add sodium hypochlorite to large private and municipal swimming pools to maintain chlorine levels. Many water sources are lakes and rivers that require addition of such chemicals as algicides to control growth of algae plus other chemicals for cleaning the water and controlling acidity levels. In wastewater treatment, the wastestream is cleaned and conditioned before the water is reintroduced into the environment. In this process, lime slurry is metered to control acidity level and polymers, coagulant aids, and ferric chloride for cleaning and conditioning.

Metering pumps are precision instruments and are used to feed accurately a predetermined volume of liquid into a process or system. They also function to pump, or convey, a liquid from one point to another. Although they are positive-displacement type pumps, they do contain special adaptations that are designed primarily to transfer liquid at an accurately controlled rate.

Metering pumps fall into four basic types, defined by the method used to seal the liquid end of the pump from the power end and prevent leakage and pumping inaccuracies:

- piston packed seal
- plunger, gland packed seal
- mechanical diaphragm seal
- hydraulic diaphragm seal

The power end of the metering pump is common to all four types, with various designs used to generate the reciprocating movement required to power the liquid end.[34]

Because of our constant demand for new and improved products and our constantly expanding knowledge in chemistry and water/wastewater treatment, there is an ever-increasing demand for precision metering of fluids. Metering pumps provide the precision metering that is needed. Will we ever develop a metering pump to match the capabilities of that human metering pump on which we all depend? Don't hold your breath waiting; the jury is still out on that one.

11.3 ROTARY PUMPS

Positive-displacement *rotary pumps* provide pumping action by the relative movement between rotating elements of the pump and stationary elements of the pump. Their rotary motion distinguishes them from reciprocating positive-displacement pumps, in which the main motion of moving elements is reciprocating.[35] Rotary pumps are primarily used as a source of fluid power in hydraulic systems. A few types are in common use in waterworks operations.

TABLE 11.1. Classes of Rotary Pumps.

Rotor Type	Pump Type
Single rotor	Vane
	Piston
	Flexible member
	Screw
	Progressive cavity
Multiple rotor	Gear
	Lobe
	Circumferential piston
	Screw

[34]From McCabe, R. E. et al., *Metering Pump Handbook.* New York: Industrial Press, Inc., p. 8, 1984.

[35]Little, C. W., Jr., *Rotary Pumps. In Pump Handbook,* Karassik, I. J. et al. (eds.). New York: McGraw-Hill, pp. 370, 1976.

Rotary pumps constitute a large class, used for relatively low flow and moderate pressures. Certain configurations, however, especially when used in hydraulic systems, may develop several thousand pounds per square inch. Table 11.1 lists the many configurations possible.

While any of the rotary pump types listed in Table 11.1 may find applications in water/wastewater treatment, this handbook focuses on the progressive-cavity pump, because of its ability to handle a wide range of fluids, from clear water to viscous solutions such as thick sludge slurries.

11.3.1 PROGRESSIVE-CAVITY PUMP

The *progressive-cavity pump* is composed of five main parts—the housing, the stator, the rotor, the connecting rod, and the drive shaft (see Figure 11.3). The rotor is usually made of either chrome-plated tool steel or stainless steel. The stator is made from natural rubber or other materials. The material selected for the rotor and stator is determined by the application. The progressive-cavity pump can achieve up to 2000 psi, depending on pump length.

In operation, liquid travels in the spaces between the rotor and the flexible stator. The rotor revolves rapidly, and capacity is directly propor-

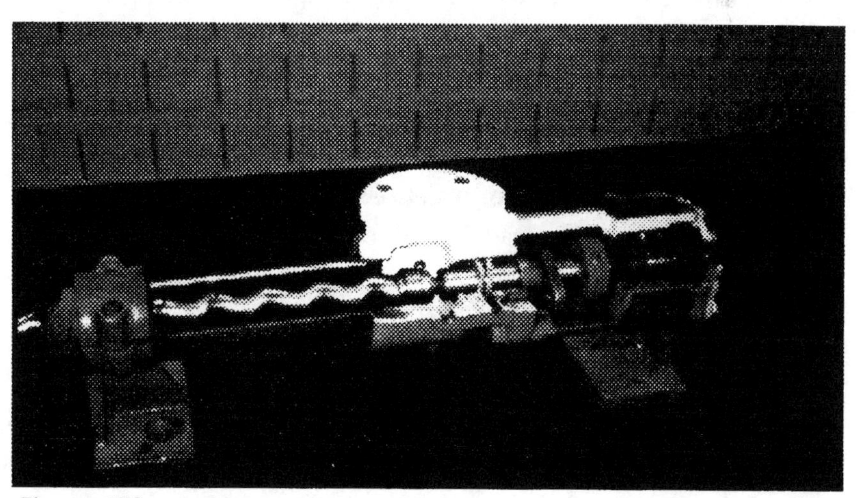

Figure 11.3
Progressive-cavity pump.

tional to rotor speed and pump size; and it produces a non-pulsing flow. No valves are associated with the pumping, but packing is needed to prevent liquid from leaking out of the pump and up the shaft. Rotor and stator are always in contact (the liquid lubricates them) as they move.[36]

IMPORTANT

Important Point: *A progressive-cavity pump should never be allowed to run dry, because it will quickly burn up and the stator will fail.*

11.4 SPECIAL-PURPOSE PUMPS

There are several special-purpose positive-displacement pumps. This handbook concentrates on peristaltic pump.

11.4.1 PERISTALTIC PUMPS

Peristaltic pumps (sometimes called tubing pumps) use a series of rollers to compress plastic tubing to move the liquid through the tubing. A rotary gear turns the rollers at a constant speed to meter the flow. Peristaltic pumps are mainly used as chemical feed pumps.

The flow rate is adjusted by changing the speed the roller-gear rotates (to push the waves faster) or by changing the size of the tubing (so there is more liquid in each wave). As long as the right type of tubing is used, peristaltic pumps can operate at discharge pressures up to 100 psi. Note that the tubing must be resistant to deterioration from the chemical being pumped. The principal item of maintenance is the periodic replacement of the tubing in the pump head. There are no check valves or diaphragms in this type of pump.

REFERENCES

AWWA, *Water Transmission and Distribution,* 2nd ed. Denver: American Water Works Association, 1996.

[36]Hauser, B. A., *Practical Hydraulics Handbook,* 2nd ed. Boca Raton, FL: Lewis Publishers, pp. 162–163, 1996.

Hauser, B. A., *Practical Hydraulics Handbook,* 2nd ed. Boca Raton, FL: Lewis Publishers, 1996.

Little, C. W., Jr., *Rotary Pumps.* In *Pump Handbook,* Karassik, I. J. et al. (eds). New York: McGraw-Hill, 1976.

McCabe, R. E., Lanckton, P. G., and Dwyer, W. V., *Metering Pump Handbook.* New York: Industrial Press, Inc., 1984.

Wahren, U. *Practical Introduction to Pumping Technology.* Houston: Gulf Publishing Company, 1997.

Self-Test

11.1 ______________ pumps move fluid by trapping a portion of fluid between one or more moving elements and a fixed casing.

11.2 Flexible tube or liner pumps, also known as ______________ pumps, are sometimes used for metering purposes.

11.3 Most metering pumps are classified as ______________ pumps or ______________ pumps.

11.4 In the peristaltic pump, fluid moves through a flexible ______ compressed by ______________.

11.5 In the progressive-cavity pump, fluid is forced toward the discharge port by the rotation of the ______________.

Final Review Examination

The answers for the comprehensive examination are contained in Appendix B.

12.1 Applications in which chemicals must be metered under high pressure require high-powered ___________ pumps.

12.2 _______________ materials resist any flow-producing force.

12.3 What type of pump is usually used for pumping high-viscosity materials?

12.4 High-powered positive-displacement pumps are used to pump chemicals that are under ________ pressure.

12.5 _______ viscosity materials are thick.

12.6 When the _______ of a pump impeller is above the level of the pumped fluid, the condition is called suction lift.

12.7 When a pump is not running, conditions are referred to as ________; when a pump is running, the conditions are __________.

12.8 With the pump _____________, the difference in elevation between the suction and discharge liquid levels is called static head.

12.9 Velocity head is expressed mathematically as _____________.

12.10 The sum of total static head, head loss, and dynamic head is called _____________.

12.11 Brake horsepower is greater than liquid horsepower by a factor representing the ___________ of the pump.

12.12 What are the three basic types of curves used for centrifugal pumps?

12.13 The liquid used to rate pump capacity is __________.

12.14 Because of the reduced amount of air pressure at high altitudes, less ____________ is available for the pump.

12.15 With the pump shut off, the difference between the suction and discharge liquid levels is called ____________.

12.16 The horsepower input to a pump is called ______________.

12.17 The ____________ and ____________ must be known to calculate brake horsepower.

12.18 ______________ is the largest single contributing factor to the reduction of pressure at a pump impeller.

12.19 The operation of a centrifugal pump is based on ____________.

12.20 The casing of a pump encloses the pump impeller, the shaft, and the ____________.

12.21 The __________ is the part of the pump that supplies energy to the fluid.

12.22 In the close-coupled pump, the impeller is mounted _________ on the motor shaft.

12.23 The impeller does not cup the water it is pumping, but slides through the fluid and __________ it.

12.24 A physical separation between the high and low pressure sides of a pump is maintained by ____________.

12.25 If wearing rings are used only on the volute case, we must replace the _______ and __________ at the same time.

12.26 Which part of the end-suction pump directs water flow into and out of the pump?

12.27 The function of the pump's impeller is to ______________.

12.28 The ______________ pump has no bearings.

12.29 The amount of pressure in a lineshaft turbine pump can be increased by adding ______________.

12.30 The ______________ propeller pump moves water only by the propelling or lifting action of the blades on the fluid.

12.31 The ______________ propeller moves water partly by centrifugal force.

12.32 Rotary pumps move fluid by trapping a portion of fluid between one or more ______________ elements and a ______________ casing.

12.33 Most metering pumps are ______________ pumps.

12.34 Flexible tube or liner pumps, also known as ______________ pumps, are sometimes used for metering purposes.

12.35 The controlled-volume, proportioning, and chemical reagent injection pumps are all ______________ metering pumps.

12.36 Most metering pumps are classified as ______________ pumps or ______________ pumps.

12.37 In the peristaltic pump, fluid moves through a flexible ______________ compressed by ______________.

12.38 In the progressive-cavity pump, fluid is forced toward the discharge port by the rotation of the ______________.

12.39 The capacity of a peristaltic pump is determined by the ______________.

12.40 Sealing devices prevent fluid ______________ along the driving shaft.

12.41 Many solid packed stuffing boxes have a ______________ that can be adjusted to reduce the amount of leakage.

12.42 In a circulating stuffing box system, the sealing water ___________ the fluid with the pump and also ___________ the packing and the shaft.

12.43 Packing should be replaced when leakage can't be controlled by tightening the ___________.

12.44 Which provides a better fluid seal, packing or a mechanical seal?

12.45 Pump ___________ support the shaft and reduce the amount of friction between the shaft and pump frame.

12.46 The two work kinds of ___________ bearings are ball and roller bearings.

12.47 What kind of lubricant is commonly used for light to moderate duty in high-speed pumps?

12.48 Grease is used as a lubricant for ___________ loads at low to moderate shaft speeds.

12.49 The most important part of bearing maintenance is ___________.

12.50 Before changing oil, it is a good practice to flush the bearing with ___________.

Appendix A

Answers to Chapter Self-Tests

CHAPTER 1

1.1 The tale of Archimedes crying "Eureka, Eureka" goes back to the Roman architect Vitruvius, who lived in the first century B.C. Archimedes had to find the volume of a sacred wreath (not, as is often said, a crown), allegedly made of pure gold, to determine whether the goldsmith had replaced some gold with silver.

"While Archimedes was turning the problem over, he chanced to come to the place of bathing, and there, as he was sitting down in the tub, he noticed that the amount of water which flowed over the tub was equal to the amount by which his body was immersed. This showed him a means of solving the problem, and he did not delay, but in his joy leapt out of the tub and, rushing naked towards his home, he cried out with a loud voice that he had found what he sought. For as he ran he repeatedly shouted in Greek, 'Eureka, Eureka' (I have found, I have found)." [37]

CHAPTER 2

2.1 Matching Exercise

1. f
2. j

[37]Stein, S., *Archimedes: What Did He Do Besides Cry Eureka?* Washington, DC: Mathematical Association of America, p. 3, 1999.

3. w
4. a
5. z
6. p
7. x
8. b
9. k
10. s
11. y
12. c
13. l
14. g
15. u
16. d
17. v
18. i
19. o
20. h
21. n
22. q
23. m
24. t
25. r
26. e

2.2 22 psi × 2.31 ft/psi = 51 ft (rounded)

2.3 $p = w \times h$

$= 62.4 \text{ lb/ft}^3 \times 16 \text{ ft}$

$= 998.4 \text{ lb/ft}^2$ or 998.4 psf

2.4 TDH = 30 × 2.31/1.03 = 67 ft (rounded)

$$bhp = \frac{800 \times 67 \times 1.03}{3960 \times 0.70}$$

$$= 19.9 \text{ hp (rounded)}$$

2.5 9.0 whp/10 bhp × 100 = 90%

CHAPTER 3

3.1 shaft, impeller, volute case

3.2 Energy from the motor is transferred through the pump shaft to spin the impeller. The spinning impeller transfers energy to the water. As the water is thrown outward by the impeller, the volute case, by nature of its design, decreases the water's velocity, which in turn increases the pressure. The increase in pressure pushes the water through the pipes.

3.3 simple and quiet operation

self-limitation of pressure

small space requirement

3.4 not self-priming

high efficiency only over a narrow range

pump can run backwards

3.5 As the water moves around the volute case, the cross-sectional area gradually increases, without an increase in the quantity of flow. This causes the velocity head to drop, which in turn, increases the pressure head. At the toe of the volute, the sudden increase in cross-sectional area completes the transfer of velocity head to pressure head.

3.6 a. 8

b. 7

c. 4

d. 1

e.	5
f.	2
g.	3
h.	10
i.	6
j.	9

CHAPTER 4

4.1 double volute

4.2 cup, slide

4.3 wearing rings

4.4 casing

4.5 supply energy to the water

4.6 bearings

4.7 open impeller

4.8 packing gland

4.9 mechanical seal

4.10 radial

4.11 open, semi-open, closed

4.12 non-clogging closed impellers

4.13 It seals the pump at the point where the pump shaft passes through the volute case. This prevents air leakage into and water leakage out of the pump.

4.14 The bearings maintain the alignment between the rotating and stationary parts of the pump and allow moving parts to rotate easily.

4.15 to join and transfer energy from the drive unit or motor to the pump

4.16 to protect the shaft from wear caused by friction as it passes through its stuffing box/packing assembly

CHAPTER 5

5.1 factory-trained service personnel

5.2 inspect pump bearings for lubrication; turn pump shaft by hand if possible; check shaft alignment; check pump rotation

5.3 check general pump operating conditions

check seal water flow

check pump control system

check discharge volume or pressure

5.4 Open the vent valve on the top of the pump casing. Slowly open the valve on the intake of the pump. Allow the casing to fill until the vent valve overflows. Close the vent valve, and start the pump. Slowly open the discharge valve until fully open.

5.5 Turn the pump off. Close the discharge control valve, and slowly open the discharge check valve and hold it open. Turn on a second pump that operates on the same discharge line. Slowly open the discharge control valve on the pump to be backflushed. Allow backflushing to continue, then slowly close the discharge control valve on backflushed pump. Close the check valve. Place backflushed pump back in service.

CHAPTER 6

6.1 bearings

6.2 Using a feeler gauge, check for angular misalignment first. Shim the motor if needed. Next, check for parallel misalignment with a straight edge. Again, shim the motor if needed. Operate the unit until warm, and immediately check both alignments again.

6.3 Shut the pump down and lockout/tagout. Remove the packing gland, and with a packing puller, remove all old packing and lantern ring. Inspect the pump shaft sleeve for wear; replace if needed. Install new packing with cuts at 180° angles. Be sure to install lantern ring in proper location. Reinstall the packing gland, and tighten finger tight. Start pump, and allow to run. Adjust packing slowly over a period of time until desired leakage is obtained.

6.4 Shut the pump down and lockout/tagout. Close the suction and discharge control valves. Dismantle the pump, and remove the old seal. Clean the shaft sleeve, and check end play and runout. Establish the location of the new seal on the shaft, and put the pump back together again. Use a feeler gauge to do the final adjustments of the rotating element.

6.5 The equipment available to do the work and the skill and knowledge of the personnel.

CHAPTER 7

7.1 Preventive maintenance is important to keep a centrifugal pump functioning smoothly and in good repair.

CHAPTER 8

8.1 3 months

8.2 proper lubrication

8.3 flush the bearing area with solvent

8.4 shaft; friction

8.5 heavy

8.6 oil

8.7 Lubrication of pump bearings will reduce friction and improve the efficiency of the pump.

8.8 oil and grease

8.9 The plug is a relief plug and allows excess grease to leave the bearing assembly during lubrication procedures.

8.10 Remove the relief plug, and clean the opening. Clean the grease fitting. Turn on the pump, and add 4–5 strokes of grease to the bearing housing. Allow the pump to run for 5–10 minutes with the relief plug out. This allows any excess or old grease to flow out. Replace the relief plug, and record the maintenance work done.

CHAPTER 9

9.1 Troubleshooting is an art that can be turned into a skill. Nothing can substitute for experience and training in effective troubleshooting. Along with experience and training, common sense is also an important factor.

CHAPTER 10

10.1 impeller-bowl assemblies

10.2 The recessed impeller pump uses an impeller recessed into the casing. The spinning action of the impeller creates a vortex or whirlpool that increases the velocity of the liquid being pumped.

10.3 It has the ability to handle liquids with large amounts of solids or debris that could clog or damage most pump impellers.

10.4 The submersible pump is placed directly in the well/wet well. It can be a close-coupled or extended shaft centrifugal pump.

10.5 The pump maintains its prime well and doesn't have air leak problems. Also, less construction is required to operate the system.

10.6 The main use is for pumping relatively clean water such as plant effluent for service water.

CHAPTER 11

11.1 rotary

11.2 peristaltic

11.3 plunger; diaphragm

11.4 tube; rollers

11.5 rotor

Appendix B

Answers to Chapter 12—Final Review Examination

12.1 positive-displacement

12.2 high-viscosity

12.3 positive-displacement

12.4 high

12.5 high

12.6 eye

12.7 static; dynamic

12.8 shut off

12.9 $V^2/2g$

12.10 total head

12.11 efficiency

12.12 head capacity, efficiency, horsepower demand

12.13 water

12.14 suction lift

12.15 elevation head

12.16 brake horsepower

12.17 required lift

12.18 water horsepower and pump efficiency

12.19 centrifugal force

12.20 stuffing box

12.21 impeller

12.22 directly

12.23 throws

12.24 wearing rings

12.25 rings; impeller

12.26 casing

12.27 supply energy to the water (fluid)

12.28 close-coupled

12.29 bowls

12.30 axial flow

12.31 mixed flow

12.32 moving; fixed

12.33 reciprocating

12.34 peristaltic

12.35 positive-displacement

12.36 plunger; diaphragm

12.37 tube; rollers

12.38 rotor

12.39 tube diameter

12.40 leakage

12.41 packing gland

12.42 seals; cools

12.43 packing gland

12.44 mechanical seal

12.45 bearings

12.46 anti-friction

12.47 oil

12.48 heavy

12.49 proper lubrication

12.50 solvent

Index

Printed in the United States
by Baker & Taylor Publisher Services